U. Reuter: Abkürzungen und Akronyme in der Chemie

ABKÜRZUNGEN UND AKRONYME IN DER CHEMIE

U. REUTER

GIT VERLAG

© 1988 by GIT VERLAG GMBH, D-6100 Darmstadt

Alle Rechte vorbehalten, insbesondere das des öffentlichen Vortrags und der fotomechanischen Wiedergabe, auch einzelner Teile.

Satz: Druckhaus Dierichs Kassel, 3500 Kassel

Printed in Germany 1988

ISBN 3-921956-73-0

Vorwort

Die Chemie mit ihren Elementsymbolen scheint prädestiniert für die Anwendung von Kurzbezeichnungen. Unabhängig von dieser Voraussetzung ist eine Zunahme von Abkürzungen und Akronymen in allen Bereichen festzustellen.

Obwohl die internationale Union für reine und angewandte Chemie (IUPAC) versuchte diese Entwicklung im Fach zu ordnen, blieb der Erfolg aus. Die inflatorische Kürzelflut hat viele Ursachen und war deshalb weder zu verlangsamen noch aufzuhalten.

Sofern heute jemand mit Chemie umgeht, wird er mit Kurzformen konfrontiert. Durch ständigen Gebrauch, weiß er mit denjenigen in seinem Arbeitsbereich umzugehen.

In der Praxis werden aber Fachgrenzen ständig mit der Folge überschritten, neuen Kurzbezeichnungen zu begegnen. Für diesen Zweck benötigt man eine Nachschlagmöglichkeit, zumal gleiche Abkürzungen verschiedene Bedeutung haben können.

Durch meine Tätigkeit an einer Universität mit vielen interdisziplinären Kontakten erkannte ich den Bedarf an einem solchen Projekt. Diese Sammlung entstand über Jahre beim schwerpunktmäßigen Lesen analytisch-chemischer Literatur. Erst eine Mehrfachnennung führte zur Aufnahme der jeweiligen Kurzform.

Die Sammlung ist bewußt einfach aufgebaut und enthält nach alphabetischer Ordnung in vier Rubriken alle Informationen.

In der ersten Spalte stehen Abkürzung bzw. Akronym, in der nächsten ggf. benutzte Alternativen. Die Bedeutung der Kurzform gibt die Spalte drei wieder. In der letzten findet sich die Zuordnung des Kürzels zu Bereichen.

Ich verdanke diese Publikation vielen Kollegen, die mich zu dieser Arbeit ermunterten. An dieser Stelle möchte ich Herrn Dr. Buchert für die motivative Unterstützung und Herrn G. Schlunck für die technische Hilfe danken, dem GIT VERLAG für die bereitwillige Umsetzung des Vorhabens.

Abkürzung Akromym / abbreviation akronym	Alternative / alternative	Bedeutung / meaning	Zuordnung / related section
AA		2-AMINO ANTHRACENE	COMP
AA		AKTIVIERUNGS ANALYSE (=ACTIVATION ANALYSIS)	METH
AA		ATOMIC ABSORPTION	INSTR
AA		ANISYLACETONE	COMP
AA	ACAC	ACETYLACETONE	COMP
AA		ABFALL AUSCHUß (VCI,D)	ORGANIS
AAA		AUTOMATED AMINOACID ANALYSES	METH
AAA		ACETOACETANILIDE	COMP
AAAF		2-(N-ACETOXYACETYLAMINO)FLUORENE	COMP
AAAS		AMERICAN ASSOCIATION FOR THE ADVANCEMENT OF SCIENCE (US)	ORGANIS
AAB		4-AMINO AZOBENZENE	COMP
AAC		AMINO ALKYL CELLULOSE	COMP
AACR		AMERICAN ASSOCIATION OF CANCER RESEARCH (US)	ORGANIS
AAD	7-AAD	7-AMINO ACTINO MYCIN	COMP
AAEU		ASTRONOMICAL ,ATMOSPHERIC EARTH AND OCEAN SCIENCES (NSF,US)	ORGANIS
AAIH		AMERICAN ACADEMY OF INDUSTRIAL HYGIENE (US)	INSTI
AAL		ATOMIC ABSORPTION INTO LINE SOURCE (HCL,AAS)	INSTR
AALA		AMERICAN ASSOCIATION FOR LABORATORY ACCREDITATION (US)	ORGANIS
AAM		AUTOMATED AUGER MICROPROBE	METH
AAMO		AMERICAN ASSOCIATION OF OCCUPATIONAL MEDICINE (US)	ORGANIS
AAMX		ACETOACET-m-XYLIDE = m-ACETOACETOXYLIDIDE	COMP
AAO		ACETALDEHYDE OXIME	COMP
AAOA		ACETOACET-o-ANISIDINE	COMP
AAOC		ACETOACET-o-CHLOROANILIDE	COMP
AAOT		ACETOACET-o-TOLUIDINE	COMP
AAS		ATOMIC ABSORPTION SPECTROMETRY	METH
AAT		ALANINE-AMINO-TRANSFERASE	CLINCHEM
AATCC		AMERICAN ASSOCIATION OF TEXTILE CHEMISTS ANS COLORISTS (US)	ORGANIS
AB		ANTIBODY	CLINCHEM
AB		AZOBENZENE	COMP

Abkürzung Akromym abbreviation akronym	Alternative alternative	Bedeutung meaning	Zuordnung related section
AB-DDR		ARZNEIBUCH DER DDR	NORM
ABA		ABISCISIC ACID	COMP
ABDA		ARBEITSGEMEINSCHAFT DER BERUFSVERTRETUNGEN DEUTSCHER APOTHEKER (D)	ORGANIS
ABIQIM		ASSOGIACAO BRASILEIRA DE INDUST.QUIMICA E DE PRODUTOS DERIVATES (BRA)	ORGANIS
ABL		ALPHA-ACETYL-GAMMA-BUTYROLACTONE	COMP
ABMP		ASSOCIATION OF BASISC MANUFACTURERS OF PESTICIDES (INDIA)	ORGANIS
ABOB		N-AMINOIMINOMETHYL-MORPHOLINE CARBOXYIMIDAMIDE	COMP
ABR		ACRYLATE-BUTADIENE-RUBBER	COMP POLYM
ABS		ALKYL BENZENE SULFONATE	COMP
ABS		ACRYLONITRILE-BUTADIENE-STYRENE TERPOLYMER	MATER POLYM
ABTS		2,2'-AZINO-DI-3-ETHYL-BENZTHIAZOLINE-6-SULFONIC ACID	COMP
ABV		ARBEITSGEMEINSCHAFT BERUFSSTÄNDIGER VERSORGUNGSEINR. (D)	INSTI
ABWAG		ABWASSER ABGABEN GESETZ (D)	REG
AC		ALTERNATING CURRENT	INSTR
AC		ACETYLATED CELLULOSE	COMP
AC-ZAA		ALTERNATING MAGNETIC CURRENT ZEEMANN ATOMIC ABSORPTION	METH
ACA	7ACA	7-AMINO CEPHALO SPORANIC ACID	COMP CLINCHEM
ACAC	AA	ACETYLACETONE	COMP
ACAR		ANGULAR CORRELATION OF ANNIHILATION RADIATION	METH
ACAST		ADVISORY COMM. ON THE APPLIC.OF SCIENCE & TECHNOLOGY TO DEVELOPMENT	ORGANIS
ACC		AMINO-CYCLOPROPANE-1-CARBOXCYLIC ACID	COMP
ACC		ADMINISTRATIVE COMMITEE ON COORDINATION (UN)	INSTI
ACCFP		ASSOCIATION OF CERAMIC COLOUR AND FRIT PRODUCERS	ORGANIS
ACCL		AMERICAN COUNCIL OF COMMERCIAL LABORATORIES (US)	ORGANIS

Abkürzung Akromym abbreviation akronym	Alternative alternative	Bedeutung meaning	Zuordnung related section
ACCM		ADVISORY COMMITTEE ON GENETIC MANUPULATION (GB)	INSTI
ACE		ALTERNATIVE CHEMICAL OR ELECTRON INPACT IONIZATION (MS)	METH
ACE		AMERICAN COUNCIL ON EDUCATION (US)	INSTI
ACE		ACETYLCHOLIN ESTERASE	CLINCHEM
ACE		ANGIOTINSINE CONVERTING ENZYME	CLINCHEM
ACES		N-(2-ACETAMIDO)-2-AMINOETHANESULFONIC ACID (=N-(CARBAMOYLMETHYLTAURINE))	COMP
ACF		ADSORBING COLLOID FLOTATION	METH
ACGIH		AMERICAN CONF.OF GOVERNMENTAL INDUSTRIAL HYGENISTS (US)	CONF
ACHEMA		AUSTELLUNGEN UND TAGUNGEN DER DECHEMA (D)	CONF
ACHR	AChR	ACETYLCHOLIN RECEPTOR	CLINCHEM
ACHT	AChT	ARBEITSAUSCHUSS CHEMISCHE TERMINOLOGIE (DIN,D)	ORGANIS
ACIT		ASSOCIATION DES CHEMISTES DE L'INDUSTRIE TEXTILE (F)	ORGANIS
ACOMAR		ADVISORY COMMITEE ON OCEANIC METEROLOGICAL RESEARCH	ORGANIS
ACP		ACYL CARRIER PROTEIN	CLINCHEM
ACP		ALTERNATING CURRENT POLAROGRAPHY	METH
ACPC		1-AMINO CYCLOPENTANE CARBOXYLIC ACID	COMP
ACS		AMERICAN CHEMICAL SOCIETY (US)	ORGANIS
ACS	7-ACS,ACA	7-AMINO CEPHALOSPORANSAEURE	COMP
ACSS		AUTOMATIC CONTINUOS SPECTRUM STABILIZATION	INSTR
ACTH		ADRENOCORTICOTROPIN = ADRENOCORTICOTROPIC HORMONE	CLINCHEM
ACV		ALTERNATING CURRENT VOLTAMMETRY	METH
ACZ		ACETYLZAHL (FETTE,SEIFEN,OELE)	MATER
ADA		N-(2-ACETAMIDO)-IMINODIACETIC ACID = N-(CARBAMOYLMETHYL) IMMINO DIACETIC ACID	COMP
ADA		ADENOSINE DEAMINASE	CLINCHEM
ADA		N-(2-ACETAMIDO)IMINODIACETIC ACID	COMP
ADAC		ANALOG-DIGITAL-ANALOG-CONVERTER	ELECT
ADAM		9-ANTHRYL DIAZO METHAN	COMP

Abkürzung Akromym abbreviation akronym	Alternative alternative	Bedeutung meaning	Zuordnung related section
ADC		AZADICARBONAMIDE	COMP
ADC		AMINO DODECYL CELLULOSE	MATER
ADC		ANALOG-TO-DIGITAL-CONVERTER	INSTR ELECT
ADCA	7-ADCA	7-AMINODESACETOXYCEPHALOSPORANIC ACID	COMP
ADDC		AMMONIUM DIETHYL DITHIOCARBAMATE	COMP
ADDM		ACETYLENE DECARBOXYLICAMINOACID DIMETHYLESTER	COMP CLINCHEM
ADEC	ADDC	AMMONIUM DIETHYL DITHIOCARBAMATE (USE ALTERNATIV)	COMP REAG
ADH		ADIURTINE HORMONE (VASOPRESSIN)	CLINCHEM
ADI		ACCEPTABLE DAILY INTAKE	CLINCHEM ENVI
ADKA		ARBEITSGEMEINSCHAFT DEUTSCHER KRANKENHAUSAPOTHEKER (D)	ORGANIS
ADM		AMORPHOUS MATRIX DENSITY	MATER
ADMA		ALKYLDIMETHYLAMINE	COMP
ADMCS		ALLYLDIMETHYLCHLORSILANE	COMP REAG
ADMS	ADMS-	ALLYLDIMETHYLSILYL-LIGAND	COMP
ADOC	ADOC-	ADAMANTYL FLUOROFORMATE-LIGAND	COMP
ADP		AMMONIUM DIHYDROGEN PHOSPHATE	COMP
ADP		ADENOSINE-5'-DIPHOSPHATE	COMP CLINCHEM
ADP	4-ADP	4-AMINODIPHENYL	COMP
ADPA	4-ADPA	4-AMINODIPHENYLAMINE	COMP
ADPC		ACETONE DIPHENYL CONDENSATION	COMP POLYM
ADPR		ADENOSINE-5'- DIPHOSPHATE RIBOSE	COMP CLINCHEM
ADR		EUROPEAN AGGREMENT ON THE TRANSPORT OF DANGEROUS GOODS BY ROAD (EUR)	REG
ADTP		AMINE DIALKYLDIPHOPHATE	COMP POLYM
ADU	ADC	ANALOG-DIGITAL-UMSETZER (UMWANDLER) (PREFER ALTERNATIV)	ELECT
ADUC		ARBEITSGEM. D.PROF F.CHEM.AN UNIV. U. HOCHSCHUL.D. BRD	ORGANIS
ADW		AKADEMIE DER WISSENSCHAFTEN (DDR)	INSTI
AE		ACOUSTIC EMISSION	INSTR
AE		AUXILLARY ELECTRODE	ECHEM INSTR
AE		AMINO ETHYL CELLULOSE	MATER
AEAPS	AEPS	AUGER ELECTRON APPEARENCE POTENTIAL SPECTROSCOPY	METH
AEAPT		(N-(2-AMINOETHYL)-3-AMINOPROPYL)TRIMETHOXYSILANE	COMP

Abkürzung Akromym	Alternative	Bedeutung	Zuordnung
abbreviation akronym	alternative	meaning	related section
AEC		ASSOCIATION EUROPEENNE CHERAMIQUE (EUR)	ORGANIS
AEC	USAEC	ATOMIC ENERGY COMMISSION (US)	INSTI
AEC-NIM		ATOMIC ENERGY A - NUCLEAR INSTRUMENT MODELS	TECHN
AEDA		ASSOCIATION EUROPEENE POUR LE DROIT DE L'ALIMENTATION	ORGANIS
AEF		AUSSCHUß FÜR EINHEITEN UND FORMELGRÖßEN IM DIN (D)	INSTI
AEM		ANALYTICAL ELECTRON MICROSCOPY	METH
AEM		AUGER ELECTRON MICROSCOPY	METH
AEMA		AUGER ELECTRON MICROANALYSIS	METH
AEO		ASSOCIATION EUROPEENNE OCEANIQUE (EUR)	ORGANIS
AEP		(2-AMINOETHYL)PIPERAZINE	COMP
AEPS	AEAPS	AUGER ELECTRON APPEARANCE POTENTIAL SPECTROSCOPY	METH
AEQCT		ASOCIATION ESPANOLA DE QUIMICOS Y COLORISTAS TEXTILES (ES)	ORGANIS
AER		ALBUMIN EXCRETION RATE	CLINCHEM
AES		ATOMIC EMISSION SPECTROMETRY	METH
AES		AUGER ELECTRON SPECTROSCOPY	METH
AES-E		AUGER ELECTRON SPECTROSCOPY WITH ELECTRON EXCITATION	METH
AES-I		AUGER ELCTRON SPECTROSCOPY WITH ION EXCITATION	METH
AES-P		AUGER ELECTRON SPECTROSCOPY WITH PHOTON EXCITATION	METH
AESGP		ASSOC. EUROPEENE DES SPESCIALITES PHARMACEUTIQUE GRAND PUBLIC	ORGANIS
AET		2-(2-AMINOETHYL)-2-THIOPSEUDOUREA DIHYDROBROMIDE	COMP REAG
AETM		ANALYTICAL ELECTRON TRANSMISSION MICROSCOPY	METH
AETS		ASSOCIATION FOR THE EDUCATION OF TEACHERS IN SCIENCE (US)	ORGANIS
AF		2-AMINO FLUORENE	COMP
AFA		AMT FÜR ARBEITSSCHUTZ (HAMBURG,D)	INSTI
AFDAC		ASSOCIATION FRANCAISE DE DOCUMENTATION AUTOMATIQUE EN CHEMIE (F)	ORGANIS

Abkürzung Akromym	Alternative	Bedeutung	Zuordnung
abbreviation akronym	alternative	meaning	related section
AFDE		ASSOCIATION FRANCAISE POUR LA DEFENCE DE L'ENVIRONMEMT (F)	ORGANIS
AFF		ABERRATION FREE FOCUS	INSTR
AFI		ASSOCIAZIONE FARMACEUTICI INDUSTRIA (I)	ORGANIS
AFID	AID	ALKALI FLAME IONIZATION DETECTOR	INSTR
AFMU		TETRAFLUORO ETHYLENE TRIFLUORONITROMETHANE NITROSOPERFLUOROBUTYRIC ACID TERPOLYMER	MATER POLYM
AFNOR		ASSOCIATION FRANCAISE DE NORMALISATION (F)	NORM ORGANIS
AFP		ALPHA-1-FETOPROTEIN	CLINCHEM
AFPTA		ASOCIATION DE FABRICANTES DE PRODUCTOS TENSIOCATIVES (ES)	ORGANIS
AFR		ARBEITSGEMEINSCHAFT ZUR FOERDERUNG DER RADIONUKLIDTECHNIK (D)	ORGANIS
AFRC		AGRICULTURAL AND FOOD RESEARCH COUNCIL (GB)	INSTI
AFS		ATOMIC FLUORESCENCE SPECTROSCOPY	METH
AFSGP		ASSOCIATION FRANCAISE DE PRODUCTURES SPECIALITES GRAND PUBLIC (F)	ORGANIS
AGA		AUSSCHUSS GEFAEHRLICHE ARBEITSSTOFFE (BMAS,D)	INSTI
AGCHEMDOK		ARBEITSGEMEINSCHAFT CHEMIEDOKUMENTATION (D)	ORGANIS
AGF		ARBEITSGEMEINSCHAFT DER GROSSFORSCHUNGSEINRICHTUNGEN (D)	ORGANIS
AGIM		ASSOCIATION GENERALE DE L'INDUSTRIE DU MEDICAMENT (BEL)	ORGANIS
AGK		ARBEITSGEMEINSCHAFT KORROSION (D)	ORGANIS
AGKR		ARBEITSGEMEINSCHAFT KRISTALLOGRAPHIE (D)	ORGANIS
AGL		AUSSCHUß GASHOCHDRUCKLEITUNGEN (D)	INSTI
AGMS		ARBEITSGEMEINSCHAFT MASSENSPEKTROMETRIE (DPG,D)	ORGANIS
AGP		ALPHA GLYCOPROTEIN	CLINCHEM
AGU		ARBEITSGEMEINSCHAFT UMWELTFRAGEN (D)	ORGANIS
AGUS	AgUS	AUSSCHUß GEFÄHRLICHE UND UMWELTRELEVANTE STOFFE (VCI,D)	ORGANIS
AGV		ARBEITSGEMEINSCHAFT DER VERBRAUCHER (D)	ORGANIS

Abkürzung Akromym abbreviation akronym	Alternative alternative	Bedeutung meaning	Zuordnung related section
AHC		AMINOHEXYL CELLULOSE	COMP
AHCTL		N-ACETYL HOMOCYTEINETHIOLACTONE	COMP
AHF		ANTIHEMOPHILIC FACTOR	CLINCHEM
AHG		ANTIHEMOPHILIC GLOBULINE	CLINCHEM
AHH		ARYLHYDROCARBON HYDROXYLASE INDUCTION TEST	CLINCHEM
AHMT	AHMT-Meth	5-AMINO-3-HYDRAZINO-5-MERCAPTOL 1,2,4-TRIAZOLE	COMP
AIA		AUTOMATED IMAGE ANALYSIS	METH
AIA		AMERICAN IMPORTERS ASSOCIATION (US)	ORGANIS
AIBN		AZO-ISO-BUTYRICACID-NITRILE =2,2'AZOBISBUTYRONITRILE	COMP POLYM
AIC		ASSOCIATION INTERNATIONALE DE LA COULEUR (INT)	ORGANIS
AICA		5(4)-AMINOIMIDAZOLE-4(5)CARBOXAMIDE	COMP
AICAR	AICAR-P	5-AMINO-IMIDAZOLE-4-CARBOXAMIDE RIBOSIDE-5-PHOSPHORIC ACID	COMP
AICF		ASSOCIAZIONE ITALIANA DI CHIMICA FISICA (I)	ORGANIS
AICHE	AIChE	AMERICAN INSTITUTE OF CHEMICAL ENGINEERS (US)	ORGANIS
AICRO		ASSOCIATION OF INDEPENDENT CONTRACT RESEARCH ORGANIZATIONS (INT)	ORGANIS
AID		ARGON IONIZATION DETECTOR	INSTR
AID	AFID	ALKALI FLAME IONIZATION DETECTOR	INSTR
AID		AGENCY FOR INTERNATIONAL DEVELOPMENT (US)	INSTI
AIDS		AQUIRED IMMUNO DEFICIENCY SYNDROME	CLINCHEM
AIEA	IAEO	AGENCE INTERNATIONALE DE L'ENERGIE ATOMIQUE (INT)	INSTI
AIF		ARBEITSGEMEINSCHAFT INDUSTRIELLER FORSCHUNGSVEREINIGUNGEN (D)	ORGANIS
AIHA		ANALYTICAL GUIDES AMERICAN INDUSTRIAL HYGIENE ASSOCIATION (US)	NORM ORGANIS
AIHC		AMERICAN INDUSTRIAL HEALTH COUNCIL (US)	INSTI
AIIBP		ASSOCIATION INTERNATIONAL DE L'INDUSTRIE DE BOULLIONS ET POTAGE	ORGANIS
AIM		ADSORPTION ISOTHERMAL MEASUREMENT	INSTR
AIOPI		ASSOC. OF INFORMATION OFFICERS IN THE PHARMACEUTICAL INDUSTRY (US)	ORGANIS

Abkürzung Akromym abbreviation akronym	Alternative alternative	Bedeutung meaning	Zuordnung related section
AIP		ALUMINIUM ISOPROPOXIDE	COMP
AIR		AEROSOL IONIC REDISTRIBUTION (AAS)	METH
AIR		5-AMINOIMIDAZOLE RIBONUCLEOTIDE	COMP
AIRM		ASSOCIAZIONE DI RISONANZE MAGNETICHEE (I)	ORGANIS
AIS		ASSOCIATION INTERNATIONALE DE LA SAVONNIERIE ET DE LA DETERGENCE	ORGANIS
AISD		ASSOCIATION DES INDUSTRIES DES SAVONS ET DES DETERGENTS (F)	ORGANIS
AK		ADENYL KINASE	CLINCHEM
AKE		ARBEITSKREIS ENERGIE (DPG,D)	ORGANIS
AKF		AUSSCHUSS KLIMA FORSCHUNG (D)	ORGANIS
AKW	PAH	POLY AROMATISCHE KOHLENWASSERSTOFFE (PREFER ALTERNATIV)	COMP
AKW	KKW	ATOMKRAFTWERKE	TECHN
ALA	Ala	ALANINE	COMP
ALA-D		DELTA-AMINOLEVUBISACID DEHYDROGENASE	CLINCHEM
ALARA		AS LOW AS REASONABLE ACHIEVABLE	REG TECHN
ALE		ARBEITSGEMEINSCHAFT FÜR LEBENSMITTEL-UND ERNÄHRUNGSWISSENSCHAFT D)	ORGANIS
ALGOL		ALGORITHMIC LANGUAGE	ELECT
ALI		ANNUAL LIMIT OF INTAKE	ENVIR REG
ALMA		ANALYTICAL LABORATORY MANAGERS ASSOCIATION (US)	ORGANIS
ALS		ARBEITSKREIS LEBENSMITTELCHEMISCHER SACHVERSTÄNDIGER (D)	ORGANIS
ALTS		ARBEITSK. LEBENSMITTELHYGIENISCHER & TIERÄRZTLICHER SACHVERSTÄNDIGER	ORGANIS
ALU		ARITHMETRIC AND LOGIC UNIT	ELECT
ALU		AUSSCHUß LEBENSMITTELHYGIENE UND LEBENSMITTELÜBERWACHUNG (D)	ORGANIS
AM		AMPLITUDE MODULATION	ELECT INSTR
AM		ALLYL METHACRYLAYE	COMP POLYM
AMA		ANTI-MITOCHONDRIAL ANTIBODY	CLINCHEM
AMA		AMERICAN MEDICAL ASSOCIATION (US)	ORGANIS
AMBA		3-AMINO-4-METHOXYBENZANILIDE	COMP

abbreviation akronym	alternative	meaning	related section
AMCHA	6AMCHA	6-AMINOMETHYL-CYCLOHEXANE CARBONIC ACID	COMP
AMCP		ASSOCIATION OF MANUFACTURES OF CHROMIUM PRODUTS	ORGANIS
AMEO		3-AMINOPROPYLTRIETHOXYSILANE	COMP
AMG		ARZNEIMITTEL GESETZ (D)	REG
AMI		ASSAY OF AMITRIPTYLINE	CLINCHEM
AMICA		AUTOMATED MODULES FOR INDUSTRIAL CONTROL ANALYSIS	INSTR
AMIF		AMERICAN MEAT INSTITUTE FOUNDATION (US)	INSTI
AMMA		ACRYLONITRILE METHYL METHACRYLATE COPOLYMER	MATER POLYM
AMMO		3-(TRIMETHOXYSILYL)-PROPYLAMINE	COMP
AMNH		AMERICAN MUSEUM OF NATURAL HISTORY (US)	INSTI
AMP		ADENOSINE-5'-MONO PHOSPHATE	COMP CLINCHEM
AMP		AMMMONIUM MOLYBDATO PHOSPHATE	COMP
AMP	bis-AMP	N-BIS(HYDROXYMETHYL)-2-AMINO-2-METHYL-1-PROPANOL	COMP
AMP-S		ADENOSINE-5'-THIOMONOPHOSPHATE	COMP
AMPD		2-AMINO-2-METHYL-1,3-PROPANEDIOLE	COMP
AMPS		2-ACRYLAMIDO-2-METHYLPROPANESULFONICACID	COMP
AMPSO		N-(1,1-DIMETHYL-2-HYDROXYETHYL)-3-AMINO-2-HYDROXY-PROPANESULFONIC ACID	COMP
AMS		AMMONIUM SULFAMATE	COMP
AMSA		4'-(9-ACRIDINYLAMINO) METHANE SULFON-M-ANISIDINE	COMP
AMSEL		ARBEITSKREIS MIKRO-UND SPURENANALSE DER ELEMENTE (D)	ORGANIS
AMTCS	AMTC	AMYLTRICHLOROSILANE	COMP
AMU	amu	ATOMIC MASS UNITS	METH THEOR
AMV		ATEM MINUTEN VOLUMEN	CLINCHEM
AN		ACETONITRILE	COMP
ANA		ANTINUCLEONIC ANTIBODY	CLINCHEM
ANA	ANS	8-ANILINO NAPHTHALENE-1-SULFONIC ACID	COMP
ANL		ARGONNE NATIONAL LABORATORY (US)	INSTI
ANM		N-(4-ANILINO-1-NAPHTHYL)MALEINIMIDE	COMP REAG

Abkürzung Akromym	Alternative	Bedeutung	Zuordnung
abbreviation akronym	alternative	meaning	related section
ANM		ACRYLESTER-ACRYLNITRIL COPOLYMER MIXTURE	MATER POLYM
ANP		ATRIAL NATRIURETIC PEPTIDE	CLINCHEM
ANP		AUSSCHUß NORMENPRAXIS IM DIN (D)	ORGANIS
ANPAT		ASSOCIATION NATIONALE POUR LA PREVENTION DES ACCIDENTS DU TRAVAIL (B)	ORGANIS
ANPO		2-(NAPHTHYL-(1')5-PHENYLOXAZOLE	COMP
ANPP		4-AZIDO-2-NITROPHENYL PHOSPHATE	COMP
ANS	ANA	8-(PHENYLAMINO)-1-NAPHTHALENESULFONATE	COMP
ANSI		AMERICAN NATIONAL STANDARDS INSTITUTE (US)	INSTI NORM
ANTU		ALPHA NAPHTHYL THIO UREA	COMP REAG
AO		ANODIC OXYDATION	METH
AO		ATOMIC ORBITAL	THEOR
AOAC		ASSOCIATION OF OFFICIAL ANALYTICAL CHEMISTS (US)	ORGANIS NORM
AOCS		ASSOCIATION OIL CHEMIST SOCIETY (US)	ORGANIS
AOD	D-& L-AOD	AMINOACID OXIDASE	CLINCHEM
AOMA		AMERICAN OCCUPATIONAL MEDICAL ASSOCIATION (US)	ORGANIS
AOP	A.O.P.	21-ACETOXY PREGNENOLONE	COMP
AOX		ADSORBABLE ORGANIC HALOGEN	MATER
AP		AMINOANTIPYRIN	COMP CLINCHEM
AP		ALKALINE PHOSPHATASE	CLINCHEM
APA	6APA ,APS	6-AMINO PENICILLAN ACID	COMP
APAD		3-ACETYLPYRIDINE ADENINE DINUCLEOTIDE	COMP
APAG		ASSOCIATION PRODUCTEURS DES ACIDES GRAS (B)	ORGANIS
APAP		N-ACETYL-p-AMINOPHENOL	COMP
APC		ACETYLSALICYLICACID-PHENACETINE-COFFEIN MIXTURE	CLINCHEM
APC		ALLOPHYTOCYANINE	COMP
APCA		AIR POLLUTION CONTROL ASSOCIATION (INT)	ORGANIS
APDC		AMMONIUMPYRROLIDIN DITHIOCARBAMATE = AMMONIUM 1-PYRROLIDINECARBODITHIOTE	COMP REAG
APEA		ASSOCIATION DES PRODUCTEURS EUROPEENS D'AZOTE (EUR)	ORGANIS

Abkürzung Akromym abbreviation akronym	Alternative alternative	Bedeutung meaning	Zuordnung related section
APECS		AUGER PHOTOELECTRON COINCIDENCE SPECTROSCOPY	METH
APEP		3-AMINO-1-PHENYL-N-ETHYL-PHENANTHRIDINIUM BROMIDE	COMP
APES		ADIABATIC POTENTIAL ENERGY SURFACE	METH
APFIM		ATOM PROBE FIELD ION MICROSCOPY	METH
APG		p-AZIDOPHENYLGLYOXAL HYDRATE	COMP
APHA		AMERICAN PUBLIC HEALTH ASSOCIATION (US)	ORGANIS
API		ATMOSPHERIC PRESSURE IONIZATION (MS)	METH
API		AMERICAN PETROLEUM INSTITUTE (US)	ORGANIS
APIACM		ASSOC. OF PRINTING INK AND ARTISTS COLOURS MANUFACTORS (INT)	ORGANIS
APL		ANTIPLASMIN	CLINCHEM
APM		N-L-ALPHA-ASPARTYL-L-PHENYLALAMINE-1-METHYLESTER = ASPANTAME	COMP
APME		ASSOCIATION OF PLASTICS MANUFACTURERS IN EUROPE	ORGANIS
APMSF	p-APMSF	(p-AMIDINOPHENYL) METHYLSULFONYLFLUORIDE	COMP
APO		AZYRIDINYLPHOSPHINOXIDE	COMP
APOBO	ApoBO	APOTHEKENBETRIEBSORDNUNG (D)	REG
APP		BIS-(2,4-DIAMINOPHENYL)PHOSPHONATE	REAG COMP
APP		ACOUSTIC PHASE PLATE	INSTR
APPE		ASSOCIATION OF PETROCHEMICALS PRODUCERS IN EUROPE	ORGANIS
APPH		AUGER PEAK TO PEAK HEIGHT	METH SPECT
APRES		ANGLE RESOLVED PHOTO-ELECTRON SPECTROSCOPY	METH
APS		ATOM PROBE SPECTROSCOPY	METH
APS		APPEARANCE POTENTIAL SPECTROSCOPY	METH
APS		ADENOSINE-5'-PHOSPHATOSULFATE	COMP
APT		(3-AMINOPROPYL) TRIMETHOXYSILANE	COMP
APT		AUTOMATIC PICTURE TRANSMISSION	METH
APT	EPDM	ETHYLENE-PROPYLENE-TERPOYMER (USE ALTERNATIVE)	MATER
APT-WEFAX		AUTOMATIC PICTURE TRANSMISSION WEATHER FAXIMILE	ENVIR TECHN
APTIC		AIR POLLUTION TECHNICAL INFORMATION CENTER (US)	INSTI

Abkürzung Akromym	Alternative	Bedeutung	Zuordnung
abbreviation akronym	alternative	meaning	related section
APTP		N-(4-AZIDOPHENYLTHIO) PHTHALIMIDE	COMP
APTS		3-AMINO PROPYL TRIETHOXYSILANE	COMP REAG
APTT		ACTIVATED PARTIAL THROMBOPLASTIN TIME	CLINCHEM
APU	ApU	ADENYLYL (3'-5'-URIDINE)	COMP
APV		ARBEITSGEMEINSCHAFT PHARMAZEUTISCHE VERFAHRENSTECHNIK (D)	ORGANIS
AQS	QA	ANALYTISCHE QUALITAETS SICHERUNG	METH
AQS		AUSSCHUSS QUALITAETSSICHERUNG UND ANGEWANDTE STATISTIK (DIN,D)	ORGANIS
AR		ANALYTICAL REAGENT	COMP
AR	PCPs	AROCHLOR (TRADE MARK ABBREVIATION)	COMP
AR		ARBEITSTAETTENRICHTLINIEN (D)	REG
ARA		ABWASSER REINIGUNGSANLAGE	TECHN
ARA-CTP		CYTOSINE ARABINOSIDE-5'-TRIPHOPHATE	CLINCHEM COMP
ARC		AGRICULTURE RESEARCH COUNCIL (GB)	ORGANIS
ARG	Arg	ARGININE	COMP
ARGE-RHEIN		ARBEITSGEMEINSCHAFT DER LÄNDER ZUR REINHALTUNG DES RHEINS	INSTI
ARI		ANNUAL RADIOTOXICITY INPUT	REG
ARIA		ANNUAL RADIOTOXICITY INPUT AIR	REG
ARIES		ANGLE RESOLVED ION ELECTRON SPECTROSCOPY	METH
ARIW		ANNUAL RADIOTOXICITY INPUT WATER	REG
ARL		APPLIED RESEARCH LABORATORIES	COMPANY
ARPA		ADVANCED RESEARCH PROJECTS AGENCY (US)	INSTI
ARS		ARYLSULFATASE	CLINCHEM
ARW		ARBEITSGEMEINSCHAFT RHEINWASSERWERKE	ORGANIS
ARXPS		ANGLE-RESOLVED X-RAY PHOTOELECTRON SPECTROSCOPY	METH
AS		AMINOSAEURE	COMP
AS		ABSORPTION SPECTROSCOPY	INSTR
ASA		AMERICAN STANDARDS ASSOCIATION (US)	NORM ORGANIS
ASA		ACETYLSALICYLIC ACID	COMP
ASA		ARGININO SUCCINIC ACID	CLINCHEM COMP
ASC		p-ACETYLAMINOBENZENESULFONYL CHLORIDE	COMP
ASCH	SCHV	ASSOCIATION SUISSE DES CHIMISTES (CH)	ORGANIS

Abkürzung Akromym abbreviation akronym	Alternative alternative	Bedeutung meaning	Zuordnung related section
ASCHIMICI		ASSOCIAZIONE NATIONALE DELL'INDUSTRIA CHIMICIA (I)	ORGANIS
ASCII		AMERICAN STANDARD CODE FOR INFORMATION INTERCHANGE	ELECT
ASCVP	SVLFC	ASSOC.SUISSE DES CHIMISTES DE l'INDUS.DES VERNIS ET PEINTUREE (CH)	ORGANIS
ASE		ALKYLSULFONICACID ESTER	COMP
ASG		ARBEITSAUSCHUß SICHERHEITSTECHNISCHE GRUNDSÄTZE IM DIN (D)	ORGANIS
ASHEA	VFWL	ASSOCIATION POUR LA SAUVEGARDE DE L'HYGIENE DE L'EAU ET DE L'AIR (CH)	ORGANIS
ASI		NATO ADVANCED STUDY INSTITUTE (INT)	CONF
ASIA		AUSTRALIAN SCIENTIFIC INDUSTRY ASSOCIATION (AUS)	ORGANIS
ASK		ARBEITSMEDIZINISCHE SACHVERSTÄNDIGEN-KOMMISSION (BG-CHEMIE ,D)	INSTI
ASKI		ARBEITSGEMEINSCHAFT DER SCHWEIZERISCHEN KUNSTSTOFFINDUSTRIE (CH)	ORGANIS
ASLEEP		AUTOMATED SCANNING LOW ENERGY ELECTRON PROBE	METH
ASM		ACOUSTIC SURFACE MEASUREMENT	INSTR
ASMA		ANTI SMOOTH MUSCLE ANTIBODY	CLINCHEM
ASN	Asn	ASPARAGINE	COMP
ASPA		ALUMINIUM SULPHATE PRODUCERS ASSOCIATION (INT)	ORGANIS
ASPEC		ASSOC.POUR LA PREVENTION E L'ETUDE DE LA CONTAMINAT (F)	ORGANIS
ASPET		AMERICAN SOC.FOR PHARMACOLOGY AND EXPERIMENTAL THERAPEUTICS (US)	ORGANIS
ASR		ALKYLENE SULFIDE RUBBER	MATER POLYM
ASRA		APPLIED SCIENCE AND RESEARCH (NSF,US)	INSTI
ASS		ANTIBIOTICA-SULFONAMIDE-SENSIBILIZATION TEST	CLINCHEM
ASSPA	SGA	ASSOCIATION SUISSE POUR L'AUTOMATIQUE (CH)	ORGANIS
AST		ASPARTATE-AMINO-TRANSFERASE	CLINCHEM
ASTE		ASSOC.POUR LE DEVELOPM.DES SCIENCES ET TECHNIQUES DE L'ENVIRONEM.(F)	ORGANIS

Abkürzung Akromym / abbreviation akronym	Alternative / alternative	Bedeutung / meaning	Zuordnung / related section
ASTM		AMERICAN SOCIETY FOR TESTING AND MATERIALS (US)	ORGANIS NORM
ASU		AMTLICHE SAMMLUNG VON UNTERSUCHUNGSVERFAHREN	LIT
ASV		ANODIC STRIPPING VOLTAMMETRY	METH
ASW		ASEPTISCHE LAMINAR FLOW WERKBANK	CLINCHEM TECHN
ASW-UP		ASEPT.LAMINARFLOW WERKBANK MIT UMWELTLUEFTUNG UND PERSONENSCHUTZ	CLINCHEN TECHN
AT		ANTITHROMBIN	CLINCHEM
ATA		APE THYREO ANTIBODY	CLINCHEM
ATA	ATA's	ALKYL TRIFLUORO ACETICACID ESTER(S)	COMP
ATA		ANTHRANILAMIDE	COMP
ATA		3-AMINO-1H-1,2,4-TRIAZOLE	COMP
ATC		ETHYLTTRICHLOROSILANE	COMP
ATC		ADDITIVES TECHNICAL COMMITTEE (EUR)	INSTI
ATCC		THE AMERICAN TYPE CULTURE COLLECTION (US)	INSTI
ATD		ALKALI THERMOIONIZATION DETECTOR	INSTR
ATEE		N-ACETYL-L-TYROSINE ETHYLESTER	COMP
ATGS		L-ALANINE TRIGLYCINSULFATE	COMP INSTR
ATMI		AMERICAN TEXTILE MANUFACTURES INSTITUTE (US)	INSTI
ATP		ADENOSINE-5'-TRIPHOSPHATE	COMP
ATPS		AZOTHIOPYRIN-DISULFONIC ACID	COMP
ATR		ATTENUATED TOTAL REFLECTION	METH
ATR		ATRACYLOSIDE	COMP CLINCHEM
ATR		AUSSCHUß TECHNISCHE RICHTLINIEN TANKCONTAINER (BERAT IM BMV,D)	INSTI
ATS		AMINOPROPYL TRIETHOXYSILANE	COMP
ATSAC		ADMINISTRATOR'S TOXIC SUBSTANCES ADVISORY COMMITTEE (US)	INSTI
ATV		ABWASSERTECHNISCHE VEREINIGUNG (BDI,D)	ORGANIS
ATZ		ANILINATHIAZOLINONE	COMP REAG
AU	UR	POLYESTER URETHANE RUBBER	MATER POLYM
AUFS	aufs	ABSORPTION-UNITS FULL SCALE	INSTR
AV-KNOTEN		ATRIOVENTRIKULAR-KNOTEN	CLINCHEM
AVA		ACCELERATING VOLTAGE ALTERATION (MS)	INSTR

Abkürzung Akromym	Alternative	Bedeutung	Zuordnung
abbreviation akronym	alternative	meaning	related section
AVCA		AGRICULTURAL AND VETERINARY CHEMICAL ASSOCIATION OF AUSTRALIA (AU)	ORGANIS
AVG		ABFALLVERBRENNUNGSGESETZ (D)	REG
AVM		ATELIER DE VITRIFICATION DE MARCULE (F)	INSTI
AVM		AVERMECTINS	COMP
AVP		ARGININE VASOPRESSIN	COMP CLINCHEM
AVV		ALLGEMEINE VERWALTUNGSVORSCHRIFTEN (D)	REG
AW		ACID WASHED	MATER
AW-DMCS		ACID WASHED DIMETHYLCHLORO SILANIZED	MATER
AWA		AUSCHUSS WASSER UND ABWASSER (IM VCI ,D)	ORGANIS
AWBR		ARBEITSGEMEINSCHAFT WASSERWERKE BODENSEE RHEIN	ORGANIS
AWI		ALFRED WEGNER INSTITUT (D)	INSTI
AWIDAT		ABFALLWIRTSCHAFTSDATENBANK (D)	LIT
AWMF		ARBEITSGEM. DER WISSENSCHAFTLICHEN MEDIZINISCHEN GESELLSCHAFTEN (D)	ORGANIS
AWT		AUSSCHUß FÜR WISSENSCHAFT UND TECHNIK (EG)	INSTI
AWTF		AUSSCHUßß FÜR WISSENSCHAFTLICHE UND TECHNISCHE FORSCHUNG (EG)	INSTI
AWTID		AUSSCHUß FÜR WISSENSCHAFTL-TECHNISCHE INFORMAT. & DOKUMENTATION (EG)	INSTI

Abkürzung Akromym abbreviation akronym	Alternative alternative	Bedeutung meaning	Zuordnung related section
BA		BUTYRALDEHYDE ANILINE	COMP POLYM
BAA		N ALPHA-BENZOYL-L-ARGININEAMIDE	COMP
BAA		BRITISH AGROCHEMICALS ASSOCIATION (GB)	ORGANIS
BAC		BENZYL-DIMETHYL-ALKYLAMMONIUMCHLORID	COMP E-ANAL
BAC		BIS-ACRYL(YL)CYSTAMIDE	COMP POLYM
BAC	1,3-BAC	1,3-BIS (AMINOETHYL)CYCLOHEXANE	COMP
BACER		BIOLOGICAL AND CLIMATIC EFFECTS RESEARCH	ENVIR
BACO		1,4-DIAZABICYCLO(2.2.2)OCTANE	COMP
BAD		BERUFSGENOSSENSCHAFTLICHER ARBEITSMEDIZINISCHER DIENST (D)	ORGANIS
BAEE		N-ALPHA-BENZOYL-L-ARGININE ETHYL ESTER	COMP
BAF	B.A.F.	BASKET ASSEMBLY FACTOR	CLINCHEM
BAFF		BUNDESANSTALT FÜR FLEISCHFORSCHUNG (D)	INSTI
BAG		BUNDESAMT FUER GESUNDHEITSWESEN (CH)	INSTI
BAH		BIOLOGISCHE ANSTALT HELGOLAND (D)	INSTI
BAH		BUNDESFACHVERBAND ARZNEIMITTEL HERSTELLER E.V. (D)	ORGANIS
BAHP		BENZOYL-MONOHYDRAZONE-3-HYDRAZINO-4-BENZYL-6-PHENYLPYRIDAZINE	COMP REAG
BAK		BUNDESAPOTHEKERKAMMER (D)	INSTI
BAL	B.A.L.	BRITISH ANTI LEWISITE = 2,3-DIMERCAPTO-1-PROPANOL	COMP
BAL		BENZALDEHYD-LIGAND	COMP
BALIS		BAYRISCHES LANDWIRTSCHAFTLICHES INFORMATIONSSYSTEM (D)	LIT
BAM		BUNDESANSTALT FUER MATERIALPRUEFUNG (D)	INSTI
BAME		N ALPHA-BENZOYL-L-ARGININE METHYL ESTER	COMP
BAN	ALPHA-BAN	ALPHA-BROMO-2'-ACETONAPHTHONE	COMP
BAN		BRITISH APPROVED NAME (GB)	REG
BANA		N-ALPHA-BENZOYL-DL-ARGININE -BETA NAPHTHYLAMIDE	COMP CLINCHEM
BANI		N ALPHA-BENZOYL-DL-ARGININE-4-NITROANILIDE	COMP
BAO		BASOL ACID OUTPUT	CLINCHEM

Abkürzung Akromym abbreviation akronym	Alternative alternative	Bedeutung meaning	Zuordnung related section
BAO		BIS-(4-AMINOPHENYL)1,3,4-OXADIAZOLE	COMP
BAP	BaP	BENZO(a)PYREN	COMP
BAP		BACTERIAL ALKALINE PHOSPHATASE	CLINCHEM
BAP		BENZYLAMINOPURINE	COMP
BAPA		N-ALPHA-BENZOYL-L-ARGININE	COMP
BAPNA		N-ALPHA-BENZOYL-DL-ARGININ-PARA-NITRIANILID	COMP CLINCHEM
BAPÖD	BApÖD	BUNDESVERBAND DER APOTHEKER IM ÖFFENTLICHEN DIENST (D)	ORGANIS
BASF		BADISCHE ANILIN UND SODA FABRIK (D)	COMPANY
BASI		BUNDESARBEITSGEMEINSCHAFT FÜR ARBEITSSICHERHEIT (DÜSSELDORF,D)	ORGANIS
BASIC		BEGINNERS ALL-PURPOSE SYMBOLIC INSTRUCTION CODE	ELECT
BASIC		BASEL INFORMATIONSZENTRUM FÜR CHEMIE (CH)	LIT
BAT		BIOLOGISCHER ARBEITSTOLERANZWERT	REG
BAV		BAYRISCHER APOTHEKER VEREIN (D)	ORGANIS
BAVC		BUNDESARBEITGEBER VERBAND CHEMIE (D)	ORGANIS
BAVCI		BUNDESARBEITGEBERVERBAND DER CHEMISCHEN INDUSTRIE (D)	ORGANIS
BAW		BUNDESANSTALT FÜR WASSERBAU (D)	INSTI
BAYLWF		BAYRISCHE LANDESANSTALT FUER WASSERFORSCHUNG (D)	INSTI
BBA		BIOLOGISCHE BUNDESANSTALT (D)	INSTI
BBBT		N,N-BIS-(p-BUTOXY-BENZYLIDINE)A,A-B-p-TOLUIDINE (PHASE)	COMP
BBC		BROWM,BOVERI COMPANIE (D)	COMPANY
BBD		7-BENZYLAMINO-4-NITRO-2,1,3 BENZOXADIAZOLE	COMP
BBD		BEAM BLANKING DEVICE	INSTR
BBF	BbF	BENZO(b)FLUANTHEN	COMP
BBM		BEAM BLANKING METHOD	METH
BBN	9-BBN	9-BONABICYCLO (3.3.1)NONANE	COMP
BBO		2,5-BIS-(4-BIPHENYL)-OXAZOLE	COMP
BBOD		2,5-BIS(4-BIPHENYLYL)-1,3,4-OXADIAZOLE	COMP
BBOT		2,5-BIS(5-TERT BUTYL-2-BENZOXAZOLYL)-THIOPHENE	COMP
BBP		BUTYLBENZYL PHTHALATE	COMP POLYM

Abkürzung Akromym abbreviation akronym	Alternative alternative	Bedeutung meaning	Zuordnung related section
BBS		BIOLOGICAL,BEHAVORIAL AND SOCIAL SCIENCES (NSF/US)	INSTI
BBS		BASLER BIOMETRISCHE SEKTION (CH)	ORGANIS
BBSP		PHENOL-3,6-DIBROMOSULFONEPHTHALEIN =PHENOL RED	COMP
BBU		BUNDESVERBAND BUERGERINITIATIVEN UMWELTSCHUTZ (D)	ORGANIS
BCA		2,2'-BICINCHONIATE	COMP REAG
BCA		N-BENZYLCYCLOPROPYLAMINE	COMP
BCB		BROMOCRESOL BLUE	COMP
BCB		5-BUTYL-1-CYCLOHEXYL BARBITURIC ACID	COMP
BCD		BINARY CODED DECIMAL	ELECT
BCDC		N-BENZYL-CINCHONIDINIUMCHLORIDE	COMP
BCEF		N,N-BIS-(2-CYANOETHYL)FORMAMIDE	COMP
BCF	BMF	BIOCONCENTRATIONFACTOR	METH
BCG		BACILLUS CALMETTE-GUERIN	CLINCHEM
BCG		BROMOCRESOL GREEN	COMP
BCHL	BChl	BACTERIO CHLOROPHYL	COMP
BCM	BMC	BEST CHLORINATING MIXTURE	COMP
BCMA		BRITISH COLOUR MAKERS ASSOCIATION (GB)	ORGANIS
BCME		BIS-CHLOROMETHYL ETHER	COMP
BCMN		BUREAU CENTRAL DES MESURES NUCLEAIRES (EG)	INSTI
BCNC		(+)-N-BENZYLCINCHODINIUM CHLORIDE	COMP
BCNU		N,N'-BIS(2-CHLOROETHYL)N-NITROSO UREA	COMP
BCP		5,5'-DIBROMO-o-CRESOLSULFON PHTHALEINE = BROMOCRESOL PURPLE	COMP CLINCHEM REAG
BCP		BUTYL CARBITOL PIPERONYLATE	COMP
BCPB		BROMOCHLOROPHENOL BLUE	COMP
BCPC		SEC-BUTYL-N-(3-CHLOROPHENYL)CARBAMATE	COMP
BCPE		1,1-BIS-(4-CHLOROPHENYL)ETHANOL (ACARIZIDE)	COMP
BCR		BUREAU COMMUNAUTAIRE DES REFERENCES (EG,CEC,EUR)	INSTI
BDAB		BENZYL-2,5-DIACETOXYBENZOATE	COMP
BDAPC		BISDIALLYLPOLYCARBONAT	COMP
BDCOH	BDC-OH	4,4'-BIS-(DIMETHYLAMINO)DIPHENYL-CARBINOL	COMP

Abkürzung Akromym abbreviation akronym	Alternative alternative	Bedeutung meaning	Zuordnung related section
BDCS	TBSCI	TETRA BUTYL DIMETHYLSILYL CHLORIDE	COMP
BDEA	t-BDEA	TERT-BUTYLDIETHANOLAMINE	COMP
BDI		BUNDESVERBAND DER DEUTSCHEN INDUSTRIE (D)	ORGANIS
BDL		BELOW DETECTION LIMIT	METH EVAL
BDL		BOLTZMANN DISTRIBUTION LAW	EVAL THEOR
BDMA		BUTANEDIOL-1,4-DIMETHYLACRYLATE	COMP POLYM
BDMA		BENZYLDIMETHYLAMINE	COMP
BDMS		BUTYLDIMETHYLSILYLETHER	COMP
BDOA		2-BIS(3',7'-DIMETHYL-2',6'OCTADIMETHYL)AMINOETHANE	COMP
BDP		BUTYL DECYL PHTHALATE	COMP POLYM
BDPA		ALPHA,GAMMA-BISDIPHENYLENE-ß-PHENYLALLYL	COMP
BDS		BUTANE DIOLE SUCCINATE	COMP
BDSA		N,O-BIS(DIMETHYL)ACETAMIDE	COMP
BDSH		BENZENE-1,3-DISULPHONYL HYDRAZIDE	COMP POLYM
BEAMOS		BEAM ADRESSED METAL OXID SEMICONDUCTOR	ELECT
BEBO		BOND ENERGY-BOND ORDER	THEOR
BEC	BEC's	BACKGROUND EQUIVALENT CONCENTRATION(S)	EVAL
BEDT-TTF		BIS(ETHYLENDITHIOLO)-TETRATHIAFULVALEN	COMP
BEE		BUREAU EUROPEEN DE L'ENVIRONNEMENT (EG)	INSTI
BEF		BUNDESAMT FUER ERNAEHRUNG UND FORSTWIRTSCHAFT (D)	INSTI
BEHP	DOP	BIS(2-ETHYLHEXYL)PHTHALATE	COMP
BER		BIT ERROR RATE	ELECT
BES		N,N-BIS(2-HYDROXYETHYL-2-AMINO-ETNANE-SULFONIC ACID	COMP
BESSY		BERLINER ELECTRONEN SPEICHER SYNCHROTON	INSTI
BET		BRUNAUER,EMMET UND TELLER (SURFACE DETERMINATION)	METH
BEUC		EUROPEAN BUREAU OF CONSUMERS (EUR)	INSTI
BFA		BUNDESFORSCHUNGSANSTALT FÜR FISCHEREI (D)	INSTI

Abkürzung Akromym / abbreviation akronym	Alternative / alternative	Bedeutung / meaning	Zuordnung / related section
BFANL		BUNDESFORSCHUNGSANSTALT F.NATURSCHUTZ & LANDSCHAFTSÖKOLOGIE (D)	INSTI
BFB		BUNDESANSTALT FUER BODENFORSCHUNG (D)	INSTI
BFE		BUNDESFORSCHUNGSANSTALT FÜR ERNÄHRUNG (D)	INSTI
BFG		BUNDESANSTALT FUER GEWAESSERKUNDE (D)	INSTI
BFS		BEAM FOIL SPECTROSCOPY	METH
BFU	BFU-E	BURST FORMING UNIT - ERYTHROCYTIC	CLINCHEM
BFU	BfU	EIDGENÖSSISCHE BERATUNGSSTELLE FÜR UNFALLVERHÜTUNG (CH)	INSTI
BFWA		BUNDESVERBAND FÜR WASSERAUFBEREITUNG (D)	ORGANIS
BG	BG-Chemie	BERUFSGENOSSENSCHAFT CHEMISCHE INDUSTRIE (D)	INSTI
BGA		BUNDESGESUNDHEITSAMT (BERLIN,D)	INSTI
BGBL		BUNDESGESTZBLATT (D)	REG
BGE		BUTYL GLYCIDYL ETHER	COMP
BGEWWI	BGewWI	BUNDESAMT FÜR GEWERBLICHE WASSERWIRTSCHAFT (D)	INSTI
BGHIP	BghiP	BENZO(ghi)PERYLEN	COMP
BGR		BUNDESANSTALT FUER GEOWISSENSCHAFTEN UND ROHSTOFFE (D)	INSTI
BGW		BUNDESVERBAND DER DEUTSCHEN GAS UND WASSERWIRTSCHAFT (D)	ORGANIS
BHA	2BHA ,3BHA	2- OR 3-TERT-BUTYL-4-HYDROXY ANISOLE (ANTIOX)	COMP
BHC	HCH,HCCH	BENZENE HEXA CHLORIDE (PREFER 1.ALTERNATIV)	COMP
BHI		BUNDESVERBAND DER ARZNEIMITTEL HERSTELLENDEN INDUSTRIE (D)	ORGANIS
BHMF		1,1'(HYDROXYMETHYL)FERROCENE	COMP
BHMF		2,5-BIS (HYDROXYMETHYL)FURAN	COMP
BHMT		BIS (HEXAMETHYLENE)TRIAMINE	COMP
BHT		2,6- OR 3,5-DI TERTBUTYL HYDROXY TOLUENE (ANTIOX)	COMP
BI-DMC		BISMUTH DIMETHYLDITHIOCARBAMATE	COMP POLYM

Abkürzung Akromym	Alternative	Bedeutung	Zuordnung
abbreviation akronym	alternative	meaning	related section
BIA		BIOLUMINESCENCE IMMUNO ASSAY	METH CLINCHEM
BIA		BERUFSGENOSSENSCHAFTLICHES INSTITUT FÜR ARBEITSSICHERHEIT (BONN,D)	ORGANIS
BIAC		BUSINESS AND INDUSTRY ADVISORY COMMITTEE (US)	INSTI
BIBRA		BRITISH INDUSTRIAL BIOLOGICAL RESEARCH ASSOCIATION (GB)	ORGABIS
BIBUQ		4,4'''-BIS-(2-BUTYLOCTYLOXY)-p-QUATERPHENYL	COMP
BIC		BIOSPECIFIC INTERACTION (AFFINITY) CHROMATOGRAPHY	METH
BICEPS		BIOINFORMATICS COLABORATIVE EUROPEAN PROGRAMM & STRATEGY	BIO
BICIN	BICINE	N,N-BIS(2-HYDROXYETHYL) GLYCINE (BUFFER)	COMP
BICT		BUNDESINSTITUT FÜR CHEMISCH-TECHNISCHE UNTERSUCHUNGEN (D)	INSTI
BIFA		BUNDESINSTITUT FÜR ARBEITSSCHUTZ (KOBLENZ,D)	INSTI
BIGA	OFIMAT	BUNDESAMT FÜR INDUSTRIE,GEWERBE UND ARBEIT (CH)	INSTI
BIIR		BROMINE ISOBUTENE ISOPRENE RUBBER	MATER POLYM
BILE	BLE	BEAM INDUCED LIGHT EMISSION	METH
BIMSCHG	BImSchG	BUNDES IMMISSIONSSCHUTZ GESETZ (D)	REG
BIMSCHV	BImSchV	BUNDESIMMISSIONSSCHUTZGESETZ STOERFALLVERORDNUNG (D)	REG
BINAP		2,2'-BIS(DIPHENYLPHOSPHINE)1,1'-BINAPHTHYL	COMP
BIOS		BRITISH INTELLIGENCE OBJECTIVES SUBCOMMITEE (GB)	INSTI
BIOSIS		BIO-SCIENCES INFORMATION SERVICES	LIT
BIPM		N-(N-(2-BENZIMIDAZOYL)-PHENYL)MALEINIMIDE	COMP
BIPM		BUREAU INTERNATIONAL DES POIDS ET MESURE (PARIS,INT)	INSTI NORM
BIS		BREMSSTRAHLUNG ISOCHROMATIC SPECTROSCOPY	METH
BIS		N,N'-METHYLENE-BIS-ACRYLAMIDE	COMP POLYM
BIS-TRIS		2,2-BIS(HYDROXYMETHYL)2,2',2''-NITILO TRIETHANOL	COMP
BISMSB	BIS-MSB	p- BIS -(O-METHYLSTYRENE)BENZENE	COMP

Abkürzung Akromym	Alternative	Bedeutung	Zuordnung
abbreviation akronym	alternative	meaning	related section
BIT		BINARY DIGIT	ELECT
BIT	ILO	BUREAU INTERNATIONAL DU TRAVAIL (GENEVE,INT)	INSTI
BIT		BUREAU INTERNATIONAL TECHNIQUE (CEFIC,INT)	INSTI
BITC	BITSC	BUREAU INTERN. TECHNIQUE DU CLORE / SOLVANTS CHLORES (CEFIC,INT)	INSTI
BIXE		BOMBARDMENT INDUCED X-RAY EMISSION	METH
BKF	BkF	BENZO(k)FLUORANTHEN	COMP
BKM		DEPARTMENT OF TYPE CULTURES-INST. OF MICROBIOLOGY (MOSCOW,SU)	INSTI
BKVO		BERUFSKRANKHEITENVERORDNUNG (D)	REG
BLAU		BUND-LÄNDER ARBEITSGEMEINSCHAFT UMWELTCHEMIKALIEN (D)	INSTI
BLE	BILE	BOMBARDMENT INDUCED LIGHT EMISSION	METH
BLG		BRENNELEMENTLAGER GORLEBEN GMBH (D)	COMP TECHN
BLIC		BUREAU DE LIEASION DE L'INDUSTRIE DU CAOUTCHOU DE LA CEE (EUR)	INSTI
BLL		BUND FÜR LEBENSMITTELRECHT UND LEBENSMITTELKUNDE (D)	ORGANIS
BLM		BILAYER MEMBRANE	ECHEM CLINCHEM
BLM		BLACK LIPID MEMBRANE	POLYM
BLO		GAMMA-BUTYROLACTONE	COMP
BLWA		BRITISH LABORATORY WARE ASSOCIATION (GB)	ORGANIS
BMAS		BUNDEMINISTERIUM FÜR ARBEIT UND SOZIALORDNUNG (D)	INSTI
BMBAU		BUNDESMINISTERIUM FUER RAUMORDNUNG,BAUWESEN UND STÄDTEBAU (D)	INSTI
BMBT		N,N-BIS-(P-METHOXY-BENZYLIDINE) A,A-BIS-p-TOLUIDINE	COMP
BMBW		BUNDESMINISTERIUM FUER BILDUNG UND WISSENSCHAFT (D)	INSTI
BMC	BCM	BALLESTER, MALINET AND CASTAUER (PERCHLORINATION)	COMP
BMC		7-BROMOMETHYL-7-METHOXY COUMAN	COMP REAG
BMC		BULK MOULDING COMPOUND	MATER POLYM
BMDC	BIDMC	BISMUTH DIMETHYL DITHIOCARBAMATE	COMP

Abkürzung Akromym	Alternative	Bedeutung	Zuordnung
abbreviation akronym	alternative	meaning	related section
BMDMCS		BROMO METHYL DIMETHYL CHLORSILANE	COMP REAG
BMDMS		BROMO METHYLDIMETHYLSILYL-REST	COMP
BMELF		BUNDESMINISTERIUM FÜR ERNÄHRUNG LANDWIRTSCHAFT UND FORSTEN (D)	INSTI
BMF	BCF	BIOMAGNIFICATION FACTOR	METH
BMFGM		BOVINE-MILK FAT GLOBULE MEMBRANE	CLINCHEM
BMFT		BUNDESMINISTERIUM FÜR FORSCHUNG UND TECHNOLOGIE (D)	INSTI
BMH		BIS-MALEIMIDO HEXANE	COMP
BMI		BUNDESMINISTERIUM DES INNEREN (D)	INSTI
BMJFG		BUNDESMINISTERIUM FUER JUGEND,FAMILIE UND GESUNDHEIT (D)	INSTI
BML		BUNDESMINISTERIUM FUER ERNAEHRUNG,LANDWIRTSCHAFT UND FORSTEN D)	INSTI
BMS		BORANE-METHYL SULFIDE COMPLEX	COMP
BMUX		BUFFERED COMMUNICATION UNIT MULTIPLEXER	ELECT
BMV		BUNDESMINISTERIUM FÜR VERKEHR (D)	INSTI
BMWI		BUNDESMINISTERIUM FUER WIRTSCHAFT (D)	INSTI
BN		BENZONITRILE	COMP
BNA		BUREAU OF NATIONAL AFFAIRS (US)	INSTI
BNAH		1-BENZYL-1,4-DIHYDRONICOTINAMIDE	COMP
BNATSCHG	BNatSchG	BUNDESNATURSCHUTZGESETZ (D)	REG
BNB		2,4,6-TRI-TERT-BUTYLNITROSOBENZENE	COMP
BNC		BAYONETT-NORM-CONNECTOR	INSTR
BNF		BRITISH NON-FERROUS METALS RESEARCH ASSOC. (GB)	ORGANIS
BNRM		BRAMMER NON-DESTRUCTIVE REFERENCE MATERIAL	MATER
BNU		BUTYL NITROSO UREA	COMP
BOA		BENZYL OCTYL ADIPATE	COMP
BOC		TERT-BUTYL-OXY-CARBONYL (REST)	COMP
BOC-ON		BOC-OXYIMMINO-2-PHENYLACETONITRIL	COMP
BOC-ONP		BOC-TERT-BUTYL-4-NITROPHENYL CARBONATE	COMP
BOC-OSU		BOC-TERT-BUTOXY-CARBAMYL OXY SUCCINIMID	COMP

Abkürzung Akromym abbreviation akronym	Alternative alternative	Bedeutung meaning	Zuordnung related section
BOC-OTCP		BOC-TERT-BUTYL-2,4,5-TRICHLOROPHENYL CARBONAT	COMP
BOD	BSB	BIOLOGICAL OXYGEN DEMAND = BIOCHEMICAL OXYGEN DEMAND	METH
BOFA		BETA ONKOFETALES ANTIGEN	CLINCHEM
BOHS		BRITISH OCCUPATION HYGIENE SOCIETY (GB)	ORGANIS
BON		BETA-OXYNAPHTHOIC ACID	COM
BOP		BENZOTRIAZOL-1-YL-OXY-TRIS-(DIMETHYLAMINO)PHOSPHONIUM HEXAFLUOROPHOSPHATE	COMP
BORAM		BLOCK ORIENTED RANDOM ACCESS MEMORY	ELECT
BP	B.P.	BRITISH PHARMACOPOEIA (GB)	NORM
BP		BENZOYLPEROXIDE	COMP
BPA		BERATENDER PROGRAMMAUSCHUSS (CEC)	INSTI
BPA		BIS-PHENOL A	COMP
BPATG		BUNDES PATENT GESETZ (D)	REG
BPB		BROMOPHENOL BLUE	COMP
BPBG		BUTYL PHTHALYL BUTYL GLYCOLATE	COMP
BPC	B.P.C.	BRITISH PHARMACEUTICAL CODEX (GB)	NORM
BPC		BONDED PHASE CHROMATOGRAPHY	METH
BPC		n-BUTYLPYRIDINIUM CHLORIDE	COMP
BPCC		2,2'-BIPYRIDINIUM CHLOROCHROMATE	COMP
BPH		2,2'-METHYLENE-BIS(4-METHYL-6-TERT-BUTYLPHENOL (ANTIOX)	COMP
BPH	BPh BPh-b	BACTERIO PHEOPHYTIN BACTERRIO PHEOPYTIN B	COMP
BPI		BUNDESVERBAND DER PHARMAZEUTISCHEN INDUSTRIE (D)	ORGANIS
BPL		BETA-PROPIOLACTAN	COMP
BPL		BRILLIANTGRUEN-PHENOLROT-LACTOSE	CLINCHEM
BPLS		BRILLIANTGRUEN-PHENOLROT-LACTOSE-SACCHAROSE	CLINCHEM
BPMP		BIS-(1-PHENYL-3-METHYLPYRAZOLON-5)	COMP REAG
BPO		2-(4-BIPHENYLYL)-5-PHENYLOXAZOLE	COMP
BPOC		2-(BIPHENYL)-PROPYL-2-YL-4-METHOXY-CARBONYLPHENYLCARBAMATE	COMP REAG
BPPM		2-DIPHENYL PHOSPHINOMETHYL PYRROLIDINE	COMP

Abkürzung Akromym abbreviation akronym	Alternative alternative	Bedeutung meaning	Zuordnung related section
BPPTB		4,5-BENZYL-1-p-CLOROPHENYL-5-PHENYL-2,4-ISODITHIOBIURET	COMP
BPR		BROMINE PYROGALLOL RED (INDICAT)	COMP
BPR		BROMOPHENOL RED	COMP
BPRS		BRIEF PSYCHIATRIC RATING SCALE	CLINCHEM
BPS		BITS PER SECOND	ELECT
BPTH		BOVINE PARATHYROID HORMONE	CLINCHEM
BPTI		BASISCHER PANKREAS TRYPSIN INHIBITOR	COMP
BR		BUTADIENE RUBBER (POLYBUTADIENE)	MATER
BRCBED	CBED	BEAM ROCKING METHOD CONVERGENT BEAM ELECTRON DIFFRACTION	METH
BRD	FRG ,D	BUNDESREPUBLIK DEUTSCHLAND	
BRDU		BROMODEOXYURIDINE	COMP CHLINCHEM
BRG		BUNDESVERBAND FÜR ROHSTOFF UND GEOWISSENSCHAFTEN (D)	ORGANIS
BRH		BUREAU OF RADIOLOGICAL HEALTH (US)	INSTI
BRILA		BRILLIANTGRUEN-GALLE-LACTOSE (NAEHRBODEN)	CLINCHEM
BROLAC		BROMTHYMOLBLAU-LACTOSE (NAEHRBODEN)	CLINCHEM
BROLACIN		BROMTHYMOLBLAU LACTASE-CYSTIN (NAEHRBODEN)	CLINCHEM
BS		BRITISH STANDARDS (GB)	NORM
BS		BACHELOR OF SCIENCE	
BS		BACKSCATTERING SPECTROMETRY	METH
BSA		N,O-BIS(TRIMETHYLSILYL) ACETAMIDE	COMP REAG
BSB	BOD	BIOLOGISCHER SAUERSTOFFBEDARF	METH
BSC		N,O-BIS (TRIMETHYLSILYL) CARBAMATE	COMP REAG
BSD		BACKSCATTERING DETECTOR	INSTR
BSE		BACKSCATTERING ELECTRONS	METH
BSF		N,N-BIS (TRIMETHYLSILYL) FORMAMIDE	COMP
BSH		BENZENE SULFONYL HYDRAZIDE	COMP POLY
BSI		BRITISH STANDARDS INSTITUTE (GB)	INSTI
BSN		BARIUM-STRONTIUM-NIOBATE (FERROMAGN)	MATER TECHN
BSOCOES		BIS- (2- (SUCCINIMIDOOXYCARBONYLOXY) ETHYL) SULFONE	COMP
BSP		BROMOPHENOL-SULFONEPHTHALEIN	COMP
BST	BST-	2-(2'-BENZOTHIAZOLYL)5-STYRYL-3-(4'-PHTHALHYDRAZIDYL)TETRAZOLIUM-LIGAND	COMP

Abkürzung Akromym / abbreviation akronym	Alternative / alternative	Bedeutung / meaning	Zuordnung / related section
BSTFA		N,O-BIS (TRIMETHYLSILYL)TRIFLUOROACETAMIDE	COMP
BSTMLU		BAYRISCHER STAATSMINISTER FÜR LANDESENTWICKLUNG UND UMWELTFRAGEN (D)	INSTI
BT		BLUE TETRAZOLIUM	COMP
BTAP		BENZYL(TRIETHYL)AMMONIUM PERMANGANATE	COM
BTB		BROMO THYMOL BLUE	COMP
BTC		BRITISH TEXTILE CONFEDERATION (GB)	ORGANIS
BTDA		3,3',4,4'-BENZOPHENONETETRACARBOXYLIC DIANHYDRIDE	COMP
BTEA		BRITISH TEXTILE EMPLOYERS ASSOCIATION (GB)	ORGANIS
BTEAC		BENZYLTRIETHYLAMMONIUM CHLORIDE	COMP
BTEE		N-BENZOYL-L-TYROSINE-ETHYL-ESTER	COMP CLINCHEM
BTFA		BIS-(TRIFLUOROACETAMIDE)	COMP
BTI		(BIS-(TRIFLUOROACETOXY) IOD) BENZENE	COMP
BTM		BETAEUBUNGSMITTEL (D)	REG COMP
BTMAHV		BETAEUBUNGSMITTEL-AUSSENHANDELSVERORDNUNG (D)	REG
BTMBINHV		BETAEUBUNGSMITTEL BINNENHANDELSVERORDNUNG (D)	REG
BTMG		BETAEUBUNGSMITTEL GESETZ (D)	REG
BTMSA		BIS(TRIMETHYLSILYL)ACETYLENE	COMP POLYM
BTMVV		BETAEUBUNGSMITTEL-VERSCHREIBUNGSVERORDNUNG (D)	REG
BTPPC		BENZYL TRIPHENYL PHOSPHONIUM CHLORIDE	COMP
BTX		BENZENE-TOLUENE-XYLENE HYDROCARBONS	COMP
BTX		BREVETOXINS	CLINCHEM COMP
BUA		BERATUNGSGREMIUM FUER UMWELTRELEVANTE ALTSTOFFE (D)	INSTI
BUCHE		BUTYRYL CHOLIN ESTERASE	CLINCHEM
BUG		BUNDES UMWELTSCHUTZ GESETZ (IN D GEFORDERT)	REG
BUND		BUNDESVERBAND UMWELT UND NATURSCHUTZ IN DEUTSCHLAND (D)	ORGANIS

Abkürzung Akromym abbreviation acronym	Alternative alternative	Bedeutung meaning	Zuordnung related section
BUS		SCHWEIZERISCHES BUNDESAMT FÜR UMWELTSCHUTZ (CH)	INSTI
BUSCAV		BIOLOGICHY USTAV CESKOSLOVENSKA AKADEMIE VED (PRAG,CSSR)	INSTI
BV		BEABSICHTIGE ZURUECKZIEHUNG EINER VORNORM (DIN ,D)	NORM
BVA		BUNDESVERBAND DER ANGESTELLTEN IN APOTHEKEN (D)	ORGANIS
BVDA		BUNDESVERBAND DEUTSCHER APOTHEKER (D)	ORGANIS
BVE		BUTYL VINYL ETHER	COMP
BVE		BUNDESVEREINIGUNG DER DEUTSCHEN ERNÄHRUNGSINDUSTRIE (D)	ORGANIS
BWH		BUREAU OF WATER HYGIENE (US)	INSTI
BZBLG		BENZIN BLEI GESETZ (D)	REG

Abkürzung Akromym abbreviation akronym	Alternative alternative	Bedeutung meaning	Zuordnung related section
CA		CHEMICAL ABSTRACT	LIT
CA		COLLISION ACTIVATION	METH
CAA		CARBOXY ARSEN AZO (INDIC)	COMP
CAA		COMPUTER AIDED ANALYSES	METH
CAB		CELLULOSE ACETATE BUTYRATE	MATER
CAC		CHEMICAL ABSTRACT CODENSATES	LIT
CACA		CANADIAN AGRICULTURAL CHEMICALS ASSOCIATION (CAN)	ORGANSI
CACP	DDP	CIS DIAMMINE DICHLORO PLATINUM	COMP
CAD		COMPUTER AIDED DISIGN	ELECT
CAD		COLLISSIONALLY ACTIVATED DISSOCIATION (MASS SPECTROMETRY)	INSTR
CAE		COMPUTER AIDED ENGINEERING	TECHN
CAF		CELLULOSE ACETAT FOIL	MATER EANAL
CAG-LIST		CARCINOGENIC ASSESSMENT GROUP LIST (OSHA , US)	NORM
CAI		COMPUTER ASSISTED INSTRUCTION	TECHN
CALA		COMMON ACUTE LEUKEMIC ANTIGENE	CLINCHEM
CAM		CHORIO ALLONTOIC MEMBRANE	CLINCHEM
CAM		COMPUTER ADDRESABLE MEMORY	ELECT
CAM		CELL-ADHESION MOLECULE	CLINCHEM
CAM	CaM CDR	CALMODOLIN	COMP CLINCHEM
CAMP	Cyc-amp	CYCLIC ADENOSINE 3',5'-MONO PHOSPHORIC ACID	COMP
CAMV		CAULIFLOWER MOSAIC VIRUS	CLINCHEM
CAN		CER (IV) AMMONIUM NITRATE	COMP
CANAS		CONFERENCE ON ANALYTICAL ATOMIC SPECTROSCOPY	CONF
CAOS		COMPUTER ASSISTED ORGANIC SYNTHESIS	METH
CAP		COLLEGUES OF AMERICAN PATHOLOGISTS (US)	ORGANIS
CAP		CLASSICAL OR CATABOLIC GENE ACTIVATING PROTEIN	CLINCHEM
CAP		CELLULOSE ACETATE PHTHALATE	MATER
CAP		CHLORACETOPHENONE	COMP
CAP		COMPUTER AIDED PLANING	TECHN
CAPD		CONTINOUS AMBULATORY PERITONEAL DIALYSIS	CLINCHEM
CAPLI2	CAP-Li2	CARBAMOYL PHOSPHATE LITHIUM	COMP

Abkürzung Akromym	Alternative	Bedeutung	Zuordnung
abbreviation akronym	alternative	meaning	related section
CAPS		3-(CYLOLHEXYL AMINO) PROPANE SULFONIC ACID (BUFFER)	COMP
CAQ		COMPUTER AIDED QUALITY ASSURENCE	METH TECHN
CAR		COMPUTER AIDED RECORDING	ELECT INSTR
CARAM		CONTENT ADDRESSABLE RANDOM ACCESS MEMORY	ELECT
CARN		CHEMICAL ABSTRACT REGISTRY NUMBER	LIT
CARS		COHERENT ANTISTOKES RAMAN SPECTROSCOPY	METH
CAS		COLLOQUIUM ATOMSPEKTROMETRISCHE SPURENANALYSE	CONF
CAS		CHEMICAL ABSTRACT SERVICE	COMPANY LIT
CASE		COOPERATIVE AWARDS IN SCIENCE AND ENGENEERING (GB)	
CASIA		CHEMICAL ABSTRACT SUBJECT INDEX ALERT (REVIEW)	LIT
CASSI		CHEMICAL ABSTRACT SERVICE SOURCE INDEX	LIT
CAST		CENTRE D'ACTUALISATION SCIENTIFIQUE ET TECHNIQUE (F)	INSTI
CAT		COMPUTER AIDED TESTING	ELECT INSTR
CAT		COMPUTER AVERAGED TRANSIENTS	ELECT
CAT		2-CHLORO-4,6-BIS(ETHYLAMINO)-s-TRIAZINE	COMP
CAT		CHLORAMPHENICOL ACETYL TRANSFERASE	CLINCHEM
CATA		COBBLE AUTOMATIC THERMO ANALYZER	INSTR
CATC		CITRAT ACID TWEEN COMBINATION	CLINCHEM
CATR		CARBOXYATRACTYLOSIDES	CLINCHEM
CB	PCB	CHLORO BIPHENYL (USE ALTERNATIVE)	COMP
CB		COOMASSIE BLUE	COMP
CB		CHEMICAL BONDED	MATER
CB		CHLOROPRENE	COMP POLYM
CB		CONVERGENT BEAM	INSTR
CB		COLLECTION OF BIOMYCETES INSTITUT OF MICROBIOLOGY (PRAG,CSSR)	INSTI
CBA	p-CBA	CARBOMETHOXY BENZALDEHYDE	COMP
CBAC		CHEMICAL-BIOLOGICAL ACTIVITIES CANADA	PROG
CBC		CARBOMETHOXY BENZENESULFONYL CHLORIDE	COMP
CBD		CANNABIDIOL	COMP
CBDS		CANNABIDIOLSAEURE	COMP

Abkürzung Akromym	Alternative	Bedeutung	Zuordnung
abbreviation akronym	alternative	meaning	related section
CBE		CARBON BLACK EXTRACT	METH
CBED		CONVERGENT BEAM ELECTRON DIFFRACTION	METH
CBI		CONFEDERATION OF BRITISH INDUSTRIES (GB)	ORGANIS
CBN		CANNABINOL	COMP
CBP		CHEMICAL BONDED PHASE	MATER
CBR		COMMUNITY BUREAU OF REFERENCE (CEC, EUR)	INSTI
CBS		BENZOTHIAZYL-2-CYCLOHEXYL SULFENEAMIDE	COMP
CBS		CENTRAALBUREAU VOOR SCHIMMELCULTURES (DELFT,NL)	INSTI
CBTP		4-CARBOXYBUTYL-TRIPHENYLPHOSPHONIUMBROMIDE	COMP
CC		CRYSTAL CONTROLLED (QUARTZ GESTEUERT)	INSTR
CC		CHEMOMETRICS IN ANALYTICAL CHEMISTRY	THEOR
CC		COLUMN CHROMATOGRAPHY	METH
CC		COUNTER CURRENT	INSTR
CCAM		COMMISSION FOR CLIMATOLOGY AND APPLICATION OF METEOROLOGY	ORGANIS
CCAP		CULTURE CENTER OF ALGAE AND PROTOZOA (CAMBRIDGE,GB)	INSTI
CCC		COUNTER CURRENT CHROMATOGRAPHY	METH
CCC		CUSTOMS COOPERATION COUNCIL (BRUSSELS,EUR)	INSTI
CCCO		COMITTEE ON CLIMATE CHANGES AND THE OCEAN	ORGANIS
CCCP		CARBON CARBON CONNECTIVITY PLOT (NMR)	METH
CCCP		CARBONYL CYANIDE-3-CHLOROPHENYLHYDRAZONE	COMP
CCD		CHARGE COUPLED DEVICE	INSTR ELECT
CCD		CLINICAL CHEMIST DIVISION (IUPAC)	ORGANIS
CCE		BETA-CARBOLINE-3-CARBOXYLATE	COMP
CCE	CEC, EG	COMMISSION DE LA COMMUNAUTE EUROPEENE	INSTI
CCEB		CULTURE COLLECTION OF ENTOMOGENEOUS BACTERIA (PRAG,CSSR)	INSTI
CCH		CYCLOHEXILIDENE CYCLOHEXANE	COMP
CCI		CHAMBRE DE COMERCE INTERNATIONALE	INSTI

Abkürzung Akromym	Alternative	Bedeutung	Zuordnung
abbreviation akronym	alternative	meaning	related section
CCK-PZ		CHOLECYSTOKININ-PANCREOZYMIN	COMP
CCM		CATION CHELATING MECHANISM	THEOR
CCM		CZECHOSLOVAK COLLECTION OF MICROORGANISMS (BRNO,CSSR)	INSTI
CCMC		CARBIDO CARBONYL METAL CLUSTER	COMP
CCMS		COMMITTEE ON THE CHALLENGES OF MODERN SOCIETY (NATO)	INSTI
CCNU		N-(2-CHLOR ETHYL(N'-CYCLOHEXYL)N-NITROSO UREA	COMP
CCO		SCIENTIFIC COMMITTEE ON THE OZON LAYER (INT)	ORGANIS
CCOHS		CANADIAN CENTRE FOR OCCUPATIONAL HEALTH AND SAFETY (CAN)	INSTI
CCP		CANADIAN CLIMATE PROGRAMME (CDN)	ENVIR
CCPA		CANADIAN CHEMICALS PRODUCERS ASSOCIATION (CAN)	ORGANIS
CCPR		CODEX COMMITEE ON PESTICIDE RESIDUE (WHO . FAO)	INSTI
CCR		COUNCIL FOR CHEMICAL RESEARCH (US)	INSTI
CCRN		COMITE CONSULTATIF DE RECHERCHE NUCLEAIRE (EG)	INSTI
CCS		CELLS CONNECTED IN SERIES	INSTR EANAL
CCSP		CONTINOUS CULTURE SIMULATION PROGRAM	CLINCHEM
CCTL		COLLECTION COUPLED TRANSISTION LOGIC	ELECT
CCTM		CENTRE DE COLLECTION DE TYPE MICROBIELLE (LILLE,F)	INSTI
CD		CIRCULAR DICHROISM	INSTR
CD		CYCLODEXTRINE	COMP
CD		COMPACT DISK	TECHN
CD50		CURRATIVE DOSIS 50 %	CLINCHEM
CDA		CYLINDRICAL DEFLECTOR ANALYZER	INSTR
CDAA		CHLORO (N-DIPROPYLENE(2)ACETAMIDE) =CHLORODIALLYLACETAMIDE	COMP
CDC		CENTRES FOR DISEASE CONTROL (US)	INSTI
CDC		COPPER DIETHYLDITHIOCARBAMATE	COMP
CDC		CYCLOHEPTAARYLOSE-DANSYL CHLORIDE COMPLEX	COMP
CDD		CHLORINATED DIBENZO-P-DIOXIN	COMP

Abkürzung Akromym abbreviation akronym	Alternative alternative	Bedeutung meaning	Zuordnung related section
CDEC		2-CHLOROALLYL-N,N- DIETHYL DITHIOCARBAMATE	COMP
CDF		CHLORO DIBENZO FURAN	COMP
CDF		CENTERED DARK FIELD	INSTR
CDM		CHLOR DIMEFORM	COMP
CDNT		4-CHLOR-3,5-DINITRO-BENZOTRIFLUORIDE	COMP
CDP		CYTIDINE-5'-DIPHOSPHATE	COMP CLINCHEM
CDQR		CONSTANT DISPLACEMENT AND QUICK RETURN	INSTR
CDR	CaM	CALCIUM-DEPENDENT REGULATOR	CLINCHEM COMP
CDRTP		CYCLODEXTRIN ROOMTEMPERATURE PHOSPHORESCENCE	INSTR
CDS		CYANODIMETHANAL SUCCINATE	COMP
CDT		CYCLODODECATRIEN	COMP
CDTA		1,2-CYCLOHEXAN DIAMINO TETRAACETIC ACID	COMP
CE		CHARGE EXCHANGE	INSTR
CE		CARLO ERBA INSTRUMENTATIONE (I)	COMPANY
CE		CARBAMAZEPINE-10,10 EPOXIDE	COMP
CE	CEC,EG	COMMUNAUTE EUROPEENE	INSTI
CEA		CARCINO EMBRYONAL ANTIGEN	CLINCHEM
CEA		COMISSARIAT A L'ENERGIE ATOMIQUE (INT)	INSTI
CEC	EG,EWG	COMISSION OF THE EUROPEAN COMMUNITIES (EG)	ORGANIS
CECD		COULSON ELECTROLYTIC CONDUCTIVITY DETECTOR	INSTR
CECE		COMBINED ELECTROLYSIS CATALYTIC EXCHANGE	TECHN
CEDRE		CENTRE D'ETUDES DE DOCUMENTAT.DE LA RECHERCHE ET D'EXPERIMENTATION (F	INSTI
CEE		COMMISION INTERN.DE REGUL.EN VUE DE L'APPROB.DE L'EQUIPE ELECTRIC	NORM
CEE	ECE	COMMISSION ECONOMIC DE NATIONS UNIES POUR L'EUROPE (UN,EUR)	INSTI
CEE-VERF		STANDARD METHODS FOR WATER QUALITY EXAMINATION FOR THE MEMBER	NORM
CEE-VERF		COUNTRIES OF THE COUNCIL FOR MUTUAL ECCONOMIC ASSISTENCE	NORM
CEEA		N-(2-CYANOETHYL)-N-ETHYLAMINE	COMP

Abkürzung Akromym	Alternative	Bedeutung	Zuordnung
abbreviation akronym	alternative	meaning	related section
CEEAS		CENTRE EUROPEEN D'ETUDES DE L'ACIDE SULFURIQUE	INSTI
CEEMT		N-(2-CYANOETHYL)-N-ETHYL-m-TOLUIDINE	COMP
CEEP		CENTRE EUROPEEN DE L'ENTERPRISE PUBLIQUE (BRUSSELS)	INSTI
CEFBC		CONTINOUS ELUTION FLAT-BED CHROMATOGRAPHY	METH
CEFIC		CONSEIL EUROPEEN DES FEDERATIONS DE L'INDUSTRIE CHEMIQUE (EUR)	ORGANIS
CEI		CENTRE D'EDUCATION	
CELS		CHARACTERISTIC ENERGY LOSS SPECTROSCOPY	METH
CEM	EM	CONVENTIONAL ELECTRON MICROSCOPY	METH
CEMA		N-(2-CYANOETHYL)-N-METHYLANILINE	COMP
CEMS		CONVERSION ELECTRON MOESSBAUER SPECTROCOPY	METH
CEN		COMMISSION EUROPEENE POUR NORME (EUR)	INSTI NORM
CENELEC		COMMISSION EUROP. POUR NORME ELECTROTECHNIQUE (EUR)	INSTI NORM
CEPA		COUPLED ELECTRON PAIR APPROXIMATION	THEOR
CEPE		COMITE EUROP.DES ASSOC.DES FABRICANTS DES PAINTURES,ENCRES D'IMPRIMES	ORGANIS
CEPEA		N-(2-HYDROXYETHYL)-N-(2-CYANOETHYL)-ANILINE	COMP
CEQ		COUNCIL ON ENVIRONMENTAL QUALITY (US)	INSTI
CERD		COMITE EUROPEEN DE RECHERCHE ET DE DEVELOPMENT (EG)	INSTI
CEREDE		CENTRE D'ETUDES ET RECHERCHE SUR EQUIPMENTS DE DEPOLLUTION DES EAUX F	INSTI
CERN		CONSEIL EUROPEEN POUR LA RECHERCHE NUCLEAIRE (GRENOBLE)	INSTI
CERTI		CABINET D'ETUDES ET DE RECHERCHES EN TOXICOLOGIE INDUSTRIE (F)	INSTI
CES		CYANO ETHYL SUCROSE	COMP
CES		CENTRE EUROPEEN DES SILCONES (BRUSSELS,EUR)	INSTI
CESFS	CES-FS	CONSTANT ENERGY SYNCHRONOMY FLUORESCENCE SPECTROMETRY	METH

Abkürzung Akromym abbreviation akronym	Alternative alternative	Bedeutung meaning	Zuordnung related section
CESIO		COMMITEE OF ORGANIC SURFACTANTS AND THEIR INTERMEDIATES (EUR)	ORGANIS
CETIS		CENTRE EUROPEEN POUR LE TRAITMENT D'INFORMATION SCIENTIFIQUE (EG)	INSTI
CEZ		CEFAZOLIN	COMP
CF		CORRECTED FIELD	INSTR
CF		CRESOL FORMALDEHYDE RESINS	MATER POLYM
CF		5(6)-CARBOXYFLUORESCEIN	COMP
CF		CRYOGENIC FOCUSSING	METH
CFA	SFA	CONTIUOS FLOW ANALYSIS	METH
CFA		CROSS FLOW ANALYZER	INSTR
CFB		CIRCULATING FLUID BED	METH TECHN
CFBC		CERCLE FRANCAIS DE BIOLOGUE CELLULAIRE	ORGANIS
CFDE		CENTRE DE FERMATION ET DE DOCUMENTATION SUR L'ENVIRONMENT (PARIS,F)	INSTI
CFE	CFM,PCTFE	POLY MONOCHLORO TRIFLUORO ETHYLENE	MATER POLYM
CFM	CFE,PCTFE	POLY MONOCHLORO TRIFLUORO ETHYLENE MIXTURE	COMP POLYM
CFPD	CfPD	CALIFORNIUM-252 PLASMA DESORPTION	INSTR
CFS		CRYSTAL FIELD STABILIZATION	THEOR
CFTA		COSMETICS,FRAGRANCES,TOILETRIES ASSOCIOATION (US)	ORGANIS
CGA		CARBONYL GROUP ACTIVATION	COMP
CGMP	c-GMP	CYCLIC GUANOSINE 3',5'-MONOPHOSPHATE	COMP
CGP	HCS	CHORIONIC GROWTH HORMONE PROLACTIN	CLINCHEM
CGPM		CONFERENCE GENERAL PAR MESURE (INT)	NORM CONF
CGRA		CORPORATE GOVERNMENT REGULATORY AFFAIRS (US)	REG
CHAP		CHRONIC HAZARDS ADVISORY PANEL (US)	INSTI REG
CHAPS		3-((3-CHOLAMIDPROPYL) DIMETHYLAMMONIO) PROPANESULFONATE	COMP
CHD		CORONARY HEART DISEASE	CLINCHEM
CHE		SERUM CHOLINESTERASE	CLINCHEM
CHEMFET	ChemFET	CHEMICAL SENSITIVE FIELD EFFECT TRANSISTOR	INSTR
CHEMG		GESETZ ZUM SCHUTZ VOR GEFAEHRLICHEN CHEMIKALIEN (D)	REG

Abkürzung Akromym abbreviation akronym	Alternative alternative	Bedeutung meaning	Zuordnung related section
CHEMTREC		CHEMICAL TRANPORTATION EMERGENCY CENTER (US)	INSTI
CHES		2-(CYCLOHEXYLAMINO)ETHANE SULFONIC ACID (BUFFER)	COMP
CHESS		CHEMICAL SHIFT SELECTIVE	METH
CHESS		COMMUNITY HEALTH AND ENVIRONMENTAL SURVEILLANCE SYSTEM (US)	ENVIR
CHIPS		CHEMICAL HAZARD INFORMATION PROFILES (US)	LIT
CHO		CHINESE HAMSTER OVARY CELLS	CLINCHEM
CHP		N-CYCLOHEXYL-2-PYRROLIDONE	COMP
CHQ		5,7-DICHLORO-8-QUINOLINOL	COMP
CHT		CYCLOHEPTATRIENE	COMP
CI		CONFIGURATION INTERACTION	THEOR
CI		CHEMICAL IONIZATION	METH
CI	C.I.	COLOUR INDEX	COMP
CIA		CHEMOLUMINESCENE IMMUNO ASSAY	METH
CIA	5-CIA	5-CHLOROISATOIC ANHYDRIDE	COMP
CIA		CHEMICAL INDUSTRIES ASSOCIATION (GB)	ORGANIS
CIAT		CENTRO INTERNATIONAL DE AGRICULTURA TROPICAL (CALI,COL)	INSTI
CID		COLLISION INDUCED DESORPTION	METH
CID		CHEMIE INFORMATION UND DOKUMENTATION (BERLIN,D)	LIT
CIDAC		CANCER INFORMATION DISSEMENATION & ANALYSIS CENTER (US)	INSTI
CIDB		CHEMIE INFORMATION UND DOKUMENTATION BERLIN (GDCH ,D)	INSTI LIT
CIDEP		CHEMICALLY INDUCED MAGNETIC ELECTRON POLARIZATION	METH
CIDNP		CHEMICALLY INDUCED DYNAMIC MAGNETIC NUCLEAR POLARIZATION	METH
CIE		COMMISSION INTERNATIONAL DE L'ECLAIRAGE	ORGANIS NORM
CIE		COUNTER IMMUNO ELECTROPHORESIS	METH
CIEL		CENTRE INTERNATIONAL D'ETUDES DU LINDANE	INSTI
CIG		COLD-INSOLUBLE GLOBULIN	CLINCHEM
CIIT		CHEMICAL INDUSTRY INSTITUTE OF TOXICOLOGY (US)	INSTI

Abkürzung Akromym abbreviation akronym	Alternative alternative	Bedeutung meaning	Zuordnung related section
CIM		COMPUTER INTEGRATED MANUFACTURING	TECHN
CIMS		CHEMICAL IONIZATION MASS SPECTROMETRY	METH
CIN		CHEMICAL INDUSTRY NOTES (REVIEW)	LIT
CIN		COUMARIN	COMP CLINCHEM
CIP		CAHN, INGOLD, PRELOG SYSTEM	THEOR
CIP	CNCM	COLLECTION DES MICROORGANISMES INSTITUT PASTEUR (PARIS,F)	INSTI
CIPC		N-(3-CHLOROPHENYL) ISOPROPYLCARBAMATE	COMP
CIRC		CENTRE INTERNATIONAL DE RECHERCHE SUR LE CANCER (LYON,F)	INSTI
CIRFS		COMITE INTERNATIONAL DE LA RAYONNE ET DE FIBRES SYNTHETIQUES (PARIS)	ORGANIS
CIRN		CENTRE D'INFORMATION DES RECHERCHES SUR LES NUISANCES (PARIS,F)	INSTI
CIS		CONTACT TO INNER SOLUTON	ECHEM
CIS		CHARACTERISTIC ISOCHROMAL SPECTROSCOPY	METH
CIS		CENTRE INTERN.D'INFORMATION DE SECURITE ET D'HYGIENE DU TRAVAIL (CH)	INSTI
CISHEC		CHEMICAL INDUSTRY SAFETY AND HEALTH COUNCIL (GB)	INSTI
CISRO		COMMONWEALTH SCIENTIFIC AND INDUSTRIAL RESEARCH ORGANIZATION (INT)	ORGANIS
CISTI		CANADA INSTITUTE FOR SCIENTIFIC & TECHNICAL INFORMATION (CDN)	LIT
CITEPA		CENTRE INTERPROF.TECHNIQUE D'ETUDES DE LA POLLUTION ATMOSPHERIQUE (F)	INSTI
CIU		COLOR INTENSITY UNIT	TECHN
CK		CREATIN KINASE	CLINCHEM
CKW	CKW's	CHLOR KOHLEN WASSERSTOFFE	COMP
CL		CHEMOLUMINESCENE	METH
CL		CATHODE LUMINESCENCE	INSTR
CLC		COLUMN LIQUID CHROMATOGRAPHY	METH
CLD		CATHODE LUMINESCENCE DETECTOR	INSTR
CLDNPYR		2-CHLOR-3,5-DINITROPYRIDINE	COMP REAG
CLEC		CHIRAL LIGAND EXCHANGE CHROMATOGRAPHY	METH

Abkürzung Akromym abbreviation akronym	Alternative alternative	Bedeutung meaning	Zuordnung related section
CLIP		CORTICOTROPIN-LIKE INTERMEDIATE LOBE PEPTIDE	CLINCHEM
CLS		CONSTANT LIGHT SIGNAL	INSTR
CLS		CORE LEVEL SPECTROCOPY	METH
CLS		CHARACTERISTIC LOSS SPECTROSCOPY	METH
CLS		CLOSED LOOP STRIPPER	INSTR
CLSA		CLOSED LOOP STRIPPING ANALYSES	METH
CLUA		CHEMISCHES LANDES UNTERSUCHUNGS AMT (D)	INSTI
CLV		CASSAVA LATENT VIRUS	CLINCHEM
CM		CAPREOMYCIN	CLINCHEM
CM		CYLINRICAL MIRROR ANALYZER	INSTR
CM-C		CARBOXYL-METHYL-CELLULOSE	COMP
CMA		N-(3-CHLOR-4-METHYLPHENYL)-2-METHYL PENTANAMID	COMP
CMA		CHEMICAL MANUFACTURERS ASSOCIATION (US)	ORGANIS
CMC		CRITICAL MICELLE CONCENTRATION	CLINCHEM
CMC		CARBOXYMETHYL-CELLULOSE	COMP
CMC		1-CYCLOHEXYL-3-(2-MORPHOLINOETHYL)CARBODIIMDE	COMP
CMDMCS		(CHLOROMETHYL) DIMETHYLCHLOROSILANE	COMP
CMDP		CARBONMETHOXYMETHYLVINYLDIMETHYL PHOSPHATE	COMP
CME-CDI		N-CYCLOHEXYL-N'-(2-MORPHOLINOETHYL) CARBODIIMIN DIMETHYL p-TOLUENESULFONATE	COMP
CMF		CROSS FLOW MICROFILTRATION	TECHN
CMHEC		CARBOXYMETHYL HYDROXYETHYLCELLULOSE	COMP
CML		CHRONIC MYELOGENOUS LEUKEMIA	CLINCHEM
CMME		CHLOROMETHYL METHYLETHER	COMP
CMNT		4-CHLORO-3-NITRO-BENZOTRIFLUORID	COMP
CMOS		COMPLEMENTARY METALOXIDE SEMICONDUCTOR	ELECT
CMP		CAPACITY COUPLED MICROWAVE PLASMA	INSTR
CMP		CYTIDINE-5'-MONOPHOSPATE	COMP
CMPP	MCPP	4-CHLOR-2-METHYLENOXYPROPIONSAEURE	COMP
CMR	C-NMR	CARBON MAGNETIC RESONANCE	METH

Abkürzung Akromym	Alternative	Bedeutung	Zuordnung
abbreviation akronym	alternative	meaning	related section
CMS		CHROMATOGRAPHIC MODE SEQUENCING	METH
CMS	CMS-	CHLOROMETHYLSILYL-LIGAND	COMP
CMTMDS		1,3-BIS(CHLOROMETHYL)-11,3,3-TETRAMETHYLDISILAZANE	COMP
CMU		CARBOXY METHYL URETHANE	COMP POLYM
CMV		CYTOMEGALOVIRUS	CLINCHEM
CN		CHLORO NAPHTHALENE	COMP
CN		CARBON NUMBERS	THEOR
CNC		CONDENSATION NUCLEI COUNTER	INSTR TECHN
CNDO		COMPLETE NEGLECT OF DIFFERENTIAL OVERLAP	THEOR
CNES		CENTRE NATIONAL D'ETUDES SPATIALES (F)	INSTI
CNG		COMPRESSED NATURAL GAS	MATER TECHN
CNIC		CENTRE NATIONAL DE L'INFORMATION CHIMIQUE (F)	INSTI
CNMR	13C-NMR	13 CARBON NUCLEAR MAGNETIC RESONANCE	METH
CNOMO		CENTRE DE NORMALISATION DE L'OUTTILLAGE ET DE MACHINES OUTILS	INSTI
CNP		2',3'-CYCLONUCLEOTIDE PHOSPHOHYDROLASE	CLINCHEM
CNRS		CENTRE NATIONAL DE LA RECHERCHE SCIENTIFIQUE (F)	INSTI
CNS	ZNS	CENTRAL NERVOUS SYSTEM	CLINCHEM
CNT		CYANOTOLUENE	COMP
CNUCED	UNCTAD	CONFERENCE DES NATIONS UNIES SUR LE COMMERS ET LE DEVELOPPEMABT (UN)	CONF
CO		POLYCHLOR METHYL OXIRAN (EPICHLORHYDRINPOLYMER)	MATER POLYM
COA	CoA	COENZYME A	COMP
COA	CoA	COAGGLUTINATION	CLINCHEM METH
COBH		CHINON-OXIM-BENZOYLHYDRAZON	COMP
COBOL		COMMOM BUISINESS ORIENTED LANGUAGE	ELECT
COCOPOL		COMMITTEE "POLICY COORDINATION ENVIRONMENTAL AND HEALTH" (CEFIC,INT)	ORGANIS
COD	CSB	CHEMICAL OXYGEN DEMAND	METH
COD		1,5-CYCLO OCTA DIENE	COMP
COD.FRANC.	Cod.Franc.	CODEX FRANCAIS	NORM
CODATA		COMITTEE ON DATA FOR SCIENCE AND TECHNOLOGY	ORGANIS
COGEMA		COMPAGNIE GENERAL DES MATIERS (F)	COMPANY

Abkürzung Akromym abbreviation akronym	Alternative alternative	Bedeutung meaning	Zuordnung related section
COLIPA		COMITE DE LIAISON DU SYNDICYT EUROPEEN DE L'IND. DEL A PARFUMERIE	ORGANIS
COMAS		CONCENTRATION-MODULATED ABSORPTION SPECTROSCOPY	METH
COMT		CATECHOL-O-METHYL TRANSFERASE	CLINCHEM
CONA		CONCANAVALIN	CLINCHEM
CONSIC		CONSORZIO PER I SERVICI ALL'INDUSTRIA CHIMICA (I)	ORGANIS
CORM		AMERICAN COUNCIL F OPTICAL RADIATION MEASUREMENT	ORGANIS
COSMOS		COMPLEMENTARY SYMMETRY METAL OXID SEMICONDUCTOR	ELECT
COSPAR		COMMITTEE ON SPACE RESEARCH	ORGANIS
COST		COOP.EUROPEENE DANS LA DOMAINE DE LA RECHERCHE SCIENT.ET TECHN	ORGANIS
COSY		CORRELATED SPECTROSCOPY	METH
COT		CYCLO OCTA TETRAENE	COMP
CP		CROSS POLARIZATION (NMR)	INSTR
CP		COMPTON PROFILE	METH
CP	n-CP	CHLOROPHENOLS (n=1,2,3,4,5)	COMP
CP	Cp	CYCLOPENTADIENE	COMP
CP	6-CP	6-CHLOROPURINE	COMP
CPA		CYTIDYLYL(3'-5')ADENOSINE	CLINCHEM
CPA	4-CPA	4-CHLOROPHENOXYACETIC ACID	COMP
CPA		CADMIUM PIGMENT PRODUCERS ASSOCIATION (INT)	ORGANIS
CPA		CONSUMER'S PROTECTION AGENCY (US)	INSTI
CPAA		CHARGED PARTICLE ACTIVATION ANALYSIS	METH
CPADS		CARBOXY PYRIDINE DISULFIDE	COMP CLINCHEM
CPAS		4'-CHLOROPHENYL-2,4,5-TRICHLOROPHENYLAZOSULFID	COMP
CPBA	m-CPBA	3-CHLOROPEROXYBENZOIC ACID	COMP
CPC		COIL PLANET CENTRIFUGE	INSTR
CPCA		CANADIAN PAINT AND COATINGS ASSOCIATION (CAN)	ORGANIS
CPD		CONTACT POTENTIAL DIFFERENCE	INSTR
CPDC		CIS-DIAMMINE DICHLORO PLATINIUM	COMP
CPE		CARBON PASTE ELECTRODE	INSTR
CPED		CONVERGENT PROBE ELECTRON DIFFRACTION	METH

Abkürzung Akromym abbreviation akronym	Alternative alternative	Bedeutung meaning	Zuordnung related section
CPFA		CYCLOPROPENOIC FATTY ACID	COMP
CPG		CONTROLLED PORE GLAS	MATER
CPI		CENTRAL PATENTS INDEX	LIT
CPI		CARBON PREFERENCE INDEX	THEOR
CPI		CHEMICAL PROCESS INDUSTRIES (US)	ORGANIS
CPIB		ETHYL 4-CHLOROPHENOXY ISOBUTYRATE	COMP
CPK		CREATIN PHOSPHOKINASE	CLINCHEM
CPK		CORREY, PAULING, KOLTUN MODEL	THEOR
CPL		CIRCULARY POLARIZED LUMINESCENCE	INSTR
CPM	cpm	COUNTS PER MINUTE	METH
CPMAR	CP/MAR	CROSS POLARIZATION MAGIC ANGLE ROTATION	METH
CPMAS	CP/MAS	CROSS POLARIZATION MAGIC ANGLE SPINNING (NMR)	METH
CPMAS		CROSS-POLARIZATION-MAGNETIC-ANGLE SPINNING (NMR)	METH
CPP		CURIE POINT PYROLYSIS	METH
CPR		CHLOROPHENOL RED	COMP
CPS		CALCIUM POLYSTYRENE SULFONATE	COMP
CPS		CONTROLLED PORE SELECTIVITY	MATER
CPS	cps	COUNTS PER SECOND	METH
CPS		CARBAMYL-PHOSPHATASE SYNTHETASE	CLINCHEM
CPS		CARBON REACTIVE POLYSACCARIDE	COMP
CPSA		CONSUMER PRODUCT SAFEY ACT (US)	REG
CPSC		CONSUMER PRODUCT SAFETY COMMISSION (US)	INSTI
CPSD		(PHOTO)COUNTING POSITION SENSITIVE DETECTOR	INSTR
CPTB		CYCLOPENTOBARBITAL	COMP
CPTEO		3-CHLOROPROPYLTRIETHOXYSILANE	COMP
CPTMO		3-CHLOROPROPYLTRIMETHOXYSILANE	COMP
CPU		CENTRAL PROCESSING UNIT	ELECT
CPVC		CHLORINATED POLYVINYLCHLORIDE	MATER POLYM
CPZ		CHLORPROMAZINE	COMP
CR		CHLOROPRENE RUBBER	MATER POLYM
CRAC		CHEMICAL REGULATIONS ADVISORY COMMITTEE (CMA,US)	ORGANIS
CRD	CRT	CATHODE RAY TUBE DISPLAY	INSTR

Abkürzung Akromym	Alternative	Bedeutung	Zuordnung
abbreviation akronym	alternative	meaning	related section
CREST		COMITE DE RECHERCHE SCIENTIFIQUE ET TECHNIQUE (EG)	INSTI
CRF		CORTICO TROPIN-RELEASING FACTOR	CLINCHEM
CRH		ACTH-RELEASING HORMONE	CLINCHEM
CRI		CROSS-REACTIVE IDIOTYPE	MATER
CRIE		CROSSED RADIOIMMUNO ELECTROPHORESIS	METH
CRM		CERTIFIED REFERENCE MATERIAL (CEC BUREAU OF REFERENCE, EUR)	MATER
CRMS		CONTINOUS REPETITIVE MEASUREMENT OF SPECTRA	METH
CRO		CATHODE RAY OSCILLOSCOPE	INSTR
CRP		CLASSICAL GENE REGULATION PROTEIN	CLINCHEM
CRP		CARBON REACTIVE PROTEIN	CLINCHEM
CRT	CRD	CATHODE RAY TUBE	INSTR
CS		D-CYCLOSERINE	COMP
CS		COMPTON SCATTERING	INSTR
CSA		COLONIA STIMULATING ACTIVITY	CLINCHEM
CSA		CAMPHORSULFONIC ACID	COMP
CSA		CANADIAN STANDARDS ASSOCIATION (CAN)	ORGANIS
CSB	COD	CHEMISCHER SAUERSTOFFBEDARF	METH
CSC		COMMITE SUISSE DE LA CHIMIE (CH)	ORGANIS
CSF		CEREBROSPINAL FLUID	CLINCHEM
CSF		COLONIA STIMULATING FACTOR	CLINCHEM
CSI		CHLORO SULFONYL ISOCYANATE	COMP REAG
CSI		COLLOQUIUM SPECTROSCOPICUM	CONF
CSI		CRYSTAL STRUCTURE IMAGING	EVAL METH
CSIN		CHEMICAL SUBSTANCES INFORMATION NETWORK (US)	LIT
CSIRO	C.S.I.R.O.	COMMONWEALTH SCIENTIFIC AND INDUSTRIAL RESEARCH ORGANIZATION	ORGANIS
CSL		COINCIDENCE SITE LATTICE	METH
CSM		CHLOROSULFONATED POLYETHYLENE MIXTURE	MATER POLYM
CSMA		CHEMICAL SPECIALITIES MANUFACTURES ASSOCIATION (US)	ORGANIS
CSN		CONDUCTIVE SOLIDS NEBULIZER	INSTR
CSP		CHIRAL SEPARATIVE (STATIONARY) PHASES	MATER
CSTR		CONTINOUS-FLOW STIRRED TANK REACTOR	TECHN

Abkürzung Akromym abbreviation akronym	Alternative alternative	Bedeutung meaning	Zuordnung related section
CSV		CATHODIC STRIPPING VOLTAMMETRY	METH
CSV	CSB,COD	CHEMISCHER SAUERSTOFF VERBRAUCH (VERALTETER BEGRIFF)	METH
CT		CALCITRANZ	CLINCHEM
CT		CHARGE TRANSFER	METH
CT		COMPUTER TOMOGRAPHY	CLINCHEM
CT		CHOLERATOXIN	CLINCHEM
CT		CRYOGENIC TRAPPING	METH
CTA		CYSTINE TRYPTOPHAN AGAR	CLINCHEM
CTA		CHEMISCH TECHNISCHER ASSISTENT (D)	
CTA	CTA-	CETYL TRIMETHYL AMMONIUM CATION	COMP
CTACN		CETYLTRIMETHYLAMMONIUM CYANIDE	COMP
CTAOH		CETYLTRIMETHYLAMMONIUM HYDROXIDE	COMP
CTAX		CETYL TRIMETHYL AMMONIUM HALIDE (X=Cl, Br)	COMP
CTC		CHLORTETRACYCLINE	COMP
CTEF		COMITE TECHNIQUE EUROPEEN FLUOR (CEFIC,EUR)	ORGANIS
CTEM	TEM	CONVENTIONAL TRANSMISSION ELECTRON MICROSCOPY	METH
CTF		CONTRAST TRANSFER FUNCTION	THEOR
CTFA		THE COSMETIC TOILETRY AND FRAGRANCE ASSOC (US)	ORGANIS
CTFE		CHLORO-TRIFLUORO-ETHYLENE	MATER
CTL		COMPLEMANTARY TRANSISTOR LOGIC	ELECT
CTM		CUNTER TIMER MULTIMETER	INSTR
CTMS		CHLORO TRIMETHYL SILANE	COMP REG
CTP		CYTIDINE-5'-TRIPHOSPHATE	COMP CLINCHEM
CTP		N-CYCLOHEXYL THIO PHTHALIMIDE	COMP
CTP		CYTIDINE THYMIDINE PHOSPHATE	COMP CLINCHEM
CTT		CHLORO TRI-ISOPROPYL ORTHOTITANATE	COMP
CTZ		CEFTEZOLE	COMP
CV		COEFFIZIENT OF VARIANCE OR VARIATION	METH EVAL
CVAA	CV-AAS	COLD VAPOR ATOMIC ABSORPTION SPECTROSCOPY	METH
CVD		CHEMICAL VAPOR DEPOSITION	TECHN
CVE		CRITICAL VOLTAGE EFFECT	METH
CVF		CIRCULAR VARIABLE FILTER	INSTR
CVS		CARDIOVASCULAR SYSTEM	CLINCHEM

Abkürzung Akromym abbreviation akronym	Alternative alternative	Bedeutung meaning	Zuordnung related section
CWA		CLEAN WATER ACT (US)	REG
CY		CYCLODIEN-BIOCIDES	COMP
CY-40		CIS CHLORDANE	COMP
CY-41		TRANS CHLORDANE	COMP
CY-50		CIS NONACHLOR	COMP
CY-51		TRANS-NONACHLOR	COMP
CYAP		O,O-DIMETHYL-O-(p-CYANOPHENYL)PHOSPHOROTHIOATE	COMP
CYDTA		CYCLOHEXANE DIAMINO TETRAACETAT LIGAND	COMP
CYP		p-CYANOPHENYL ETHYL PHENYLPHOSPHORIC ACID	COMP
CYS	Cys	CYSTINE	COMP
CYSH	CySH	CYSTEINE	COMP
CZE		CAPILLARY ZONE ELECTROPHORESIS	METH

Abkürzung Akromym	Alternative	Bedeutung	Zuordnung
abbreviation akronym	alternative	meaning	related section
D	2,4 D	2,4-DICHLORPHENOXY ACETIC ACID	COMP
D		2,2'-DITHIOBENZOIC ACID	COMP
D ESTER	2,4D ESTER	2,4-DICHLORPHENOXY ACETIC ACID ESTER	COMP
DA		DOPAMINE	COMP CLINCHEM
DAA		DIACETONE ACRYLAMIDE	COMP
DAA		DIACETONE ALCOHOL	COMP
DAAB	DAB	4-DIMETHYLAMINO-AZOBENZENE	COMP
DAAD		DEUTSCHER AKADEMISCHER AUSLANDSDIENST (D)	INSTI
DAAO		D-AMINO ACID OXIDASE	CLINCHEM
DAB	DAB 9	DEUTSCHES ARZNEIBUCH NUMMER 9	NORM
DAB	DAAB	4-N,N-DIMETHYLAMINOAZOBENZENE	COMP
DAB		DIAMINOBENZIDIENE	COMP
DABAWAS		DATENBANK WASSERGEFAERDENDER STOFFE (D)	LIT
DABC	3,5-DABC	3,5-DIAMINOCHLOROBENZENE	COMP
DABCO	TED	1,4-DIAZABICYCLO(2.2.2)OCTANE	COMP
DABF	DAbF	DEUTSCHER AUSSCHUSS FUER BRENNBARE FLUESSIGKEITEN (TÜV,D)	ORGANIS
DABITC		4-(N,N-DIMETHYLAMINO-AZOBENZENE-4'-ISOTHIOCYANATE	COMP
DABS	DABS-Cl	4-(N,N-DIMETHYLAMINO)AZOBENZENE-4'-SULFONYLCHLRIDE	COMP
DABTC		N,N-DIMETHYLAMINO-AZO BENZENE-THIOCARBONYL- LIGAND	COMP
DABTH		N,N-DIMETHYL-AMINO-AZOBENZENE-THIOHYDANTOIN	COMP
DABTZ	DABTZ-	N,N-DIMETHYL AMINO AZOBENZENE THIAZOLINONE-LIGAND	COMP
DAC		DEUTSCHER ARZNEIMITTEL CODEX	NORM
DAC	DAU	DIGITAL TO ANALOG CONVERTER	INSTR
DACA	DAcA	DEUTSCHER ACETYLEN AUSSCHUß	ORGANIS
DACH		TRANS-1,2-DIAMINOCYCLOHEXANE	COMP
DACM-3		N-(7-DIMETHYLAMINO-4-METHYL-3-COUMARINYL)MALEINIMIDE	COMP REAG
DACS		DATA ACQUISITION CONTROL SYSTEM	ELECT
DACS	DACs	DIAMOND ANVIL CELLS	MATER
DAD		DIODE ARRAY DETECTOR	INSTR
DAD		DIGITAL AUDIO DISK	TECHN

Abkürzung Akromym / abbreviation akronym	Alternative / alternative	Bedeutung / meaning	Zuordnung / related section
DADA	DIPA-DCA	DIISOPROPYL AMMONIUM DICHLORO ACETATE	COMP
DADDS		4,4'-DIACETYLAMINO DIPHENYL SULFONE	COMP
DADI		DIRECT ANALSIS OF DAUGHTER IONS	METH
DADI		DIANISIDINE DI ISOCYANATE	COMP POLYM
DADPM		4,4'-DIAMINO-DIPHENYLMETHANE	COMP
DADPS	DDS	DIA DIPHENYL SULFONE	COMP
DAEC		DIETHYL AMINOETHYL CELLULOSE	COMP
DAGST		DEUTSCHER AUSSCHUSS FUER GRENZFLAECHEN AKTIVE STOFFE (DIN,D)	ORGANIS
DAGST		DEUTECHER AUSSCHUß FÜR GRENZFLÄCHENAKTIVE STOFFE	ORGANIS
DAL		DEITSCHER ARBEITSRING FÜR LÄRMBEKÄMPFUNG	ORGANIS
DAMN		DIAMINOMALEONITRILE	COMP
DAMO		N-AMINOETHYLAMINOPROPYL-TRIMETHOXYSILANE	COMP
DANS	DANSYL	DIMETHYLAMINO-1-NAPHTHALENE SULFONYL-LIGAND	COMP
DAP	3,4-DAP	3,4-DIAMINOPYRIN	COMP CLINCHEM
DAP		DIPHENYL AMINE ANILINE PHOSPHATE	COMP
DAP		DIAMINO PIMELINSAEURE	COMP
DAP		DIALLYL PHTHALATE	COMP
DAP		DIAMMONIUM PHOSPHATE	COMP
DAPA		DEUTSCHER AUSSCHUß FÜR PFLANZENSCHUTZANALYTIK	ORGANIS
DAPI		4',6-DIAMIDINO-2-PHENYL INDOLE	COMP
DAPI		DEUTSCHES ARZNEIPRÜFUNGS INSTITUT (D)	INSTI
DAPS		DIS APPEARANCE POTENTIAL SPECTROCOPY	METH
DAPT		2,4-DIAMINO-5-PHENYL THIAZOLE	COMP
DAS	D.A.S.	DEUTSCHE AUSLEGESCHRIFT (D)	REG
DAS		9,10-DIMETHOXY ANTHRACENE-2-SULFONIC ACID	COMP
DAS		4,4'DIAMINOSTILBENE-2,2'DISULFONIC ACID	COMP
DASP		DOUBLE ANTIBODY SOLID PHASE	CLINCHEM
DASP	DASp	DEUTSCHER ARBEITSKREIS FUER ANGEWANDTE SPEKTROSCOPIE (GDCH,D)	ORGANIS
DAST		(DIETHYLAMINO)SULFUR-TRIFLUORIDE	COMP REAG
DAT		DEUTSCHE ARZNEITAXE (D)	REG

Abkürzung Akromym / abbreviation akronym	Alternative / alternative	Bedeutung / meaning	Zuordnung / related section
DATC		S-2,3-DICHLOROALLYL DIISOPROPYL THIOCARBAMATE	COMP
DATD		N,N'-DIALLYL-TARTRAIC ACID DIAMIDE	COMP
DATMP		DIETHYL ALUMINIUM-2,2,6,6-TETRAMETHYL PIPERIDINE	COMP
DAU		DISKUSSIONSKREIS ANALYTIK IM UMWELTSCHUTZ (GDCH,D)	ORGANIS
DAU	DAC	DIGITAL-ANALOG UMSETZER (VERALTET)	ELECT INSTR
DB		DOUBLE BEAM	INSTR
DB	2,4DB	4(2,4-DICHLOROPHENOXY) BUTANOIC ACID	COMP
DBA		4,4'-DICHLORO BENZILIC ACID	COMP
DBA		DIBENZA(a,h)ANTHRACENE	COMP
DBB		DIBROMO BENZENE	COMP
DBBP	PBB2	DIBROMO BIPHENYL	COMP
DBC		DIBENZO-18-CROWN	COMP
DBCP		1,2-DIBROMO-3-CHLOROPROPANE	COMP
DBD		DIBROMO DULCIT =MITOLACTOL	COMP
DBDPO		DECABROMODIPHENYLOXIDE	COMP
DBE		1,2-DIBROMO ETHANE	COMP
DBED	7-DBED	DIBENZYL ETHYLENE DIAMINE	COMP
DBED	DBED-X	DIBENZYLETHYLENEDIAMINE DIPENICILLIN X=DIV.	COMP CLINCHEM
DBEEA		DIBUTOXY ETHOXY ETHYL ADIPATE	COMP POLYM
DBF		DIBENZOFURAN	COMP
DBG		DEUTSCHE BUNSEN GESELLSCHAFT FÜR PHYSIKALISCHE CHEMIE (D)	ORGANIS
DBH	4,4'DBH	BIS (4-CHLOROPHENYL)METHANAL	COMP
DBHT		DIBUTYL HYDROXY TOLUENE	COMP
DBIC		DIBUTYL INDOLO CARBAZOLE	COMP
DBM		DIBROMOMANNITOL =MITOBROMITOL	COMP
DBMIB		DIBROMO METHYL ISOPROPYL BENZOQUINONE	COMP
DBN		1,5-DIAZOBICYCLO(4.3.0)NON-5-ENE	COMP
DBN		2,6-DICHLORO BENZONITRILE	COMP
DBP	D.B.P.	DEUTSCHES BUNDESPATENT (D)	REG
DBP	DCBP	DICHLORO BENZOPHENONE (2,4' & 4,4' SUBSTITUTION)	COMP
DBP		DIBUTYL PHTHALATE	COMP
DBPC		2,6-DI-TERT-BUTYL-p-CRESOL	COMP

Abkürzung Akromym	Alternative	Bedeutung	Zuordnung
abbreviation akronym	alternative	meaning	related section
DBPPD		N,N-DI-S-BUTYL-4-PHENYLENE DIAMINE	COMP POLYM
DBS		SODIUM DOCECYL BENZENE SULFONATE	COMP
DBS		DIBUTYL SEBACATE	COMP
DBTCP		DIBUTYL TETRACHLORO PHTHALATE	COMP
DBTU		DIBUTYL THIO UREA	COMP
DBU		1,8-DIAZABICYCLO(5,4,0)UNDEC-7-ENE	COMP
DC	TLC	DÜNNSCHICHT CHROMATORGRAPHIE	METH
DC		DIFFERENTIAL CURRENT	INSTR
DC		DIFFRACTION CONTRAST	INSTR
DC-ZAA		DIRECT CURRENT ZEEMAN ATOMIC ABSORPTION	METH
DCA	4,4DCA	BIS(4-CHLOROPHENYL) ACETIC ACID	COMP
DCA		DICHLORO ACETIC ACID	COMP
DCA		DESOXY CARTANE ACETATE	COMP
DCAD	2,4-DCAD	2,4-DICHLORO BENZALDEHYDE	COMP
DCAF		1,8-BIS-(DI(CARBOXYMETHYL) AMINOMETHYL)-FLUORESCEINE	COMP
DCB		DICHLORO BUTADIENE	COMP POLYM
DCB		DICYANO BENZENE	COMP
DCB		DICHLORO BENZENE & DICHLORO BENZIDINE	COMP
DCBA	2,4-DCBA	2,4-DICHLOROBENZOIC ACID	COMP
DCBC	2,4-DCBC	2,4-DICHLOROBENZYL CHLORIDE	COMP
DCBP	DBP	4,4'-DICHLORO DIBENZOPHENONE	COMP
DCBS		BENZO THIAZYL-2-DICYCLOHEXYL SULPHENAMIDE	COMP POLYM
DCBTF		DIVERSE-DICHLORO BENZO TRIFLUORIDE	COMP
DCC		DEXTRAN COATED CHARCOAL	MATER
DCC		DIPHENYL CARBONYL CHLORIDE	COMP CLINCHEM
DCC	DCCD	DICYCLOHEXYL CARBODIIMIDE	COMP
DCCC		DROPLET COUNTER CURRENT CHROMATOGRAPHY	METH
DCCD	DCC	DICYCLO HEXYL CARBODIIMIDE	COMP
DCCP		DEPARTMENT OF CANCER CLAUSE AND PREVENTION (US)	INSTI
DCDD	PCDD2	DICHLORO-P-DIBENZODIOXIN	COMP
DCDF	PCDF2	DICHLORO DIBENZO FURAN	COMP
DCE		1,1 UND AUCH 1,2-DICHLOROETHYLENE	COMP
DCE		1,2-DICHLORETHANE	COMP

Abkürzung Akromym abbreviation akronym	Alternative alternative	Bedeutung meaning	Zuordnung related section
DCEE		DICHLORORETHYL ETHER	COMP
DCET		O,S-DICARB ETHOXY THIAMINE	COMP
DCH18C6		DICYCLOHEXYL-18-CROWN-6	COMP
DCHA		DICYCLOHEXYLAMINE	COMP
DCHBH		DICYCLOHEXYLBORANE	COMP
DCHP		DICYCLOHEXYLPHTHALATE	COMP
DCI		DESORPTION CHEMICAL IONIZATION	METH
DCI		1-(3',4'-DICHLOROPHENYL)-2-ISOPROPYL AMINOETHANOL	COMP
DCL		DEMOUNTABLE CATHODE LAMP	INSTR
DCLS		DESOXYCHOLAT-CITRAT-LACTOSE-SACCHAROSE	CLINCHEM
DCLT		DIFFERENTIAL CATHODE LUMINESCENCE TOPOGRAPHY	METH
DCMA		DRY COLOUR MANUFACTURERS ASSOCIATION (US)	ORGANIS
DCMC		DICHLOROMETA XYLENOL	COMP
DCME		ALPHA,ALPHA-DICHLORORO METHYLETHER	COMP
DCMO		2,3-DIHYDRO-5-CARBOXANILIDO-6-METHYL-1,4-OXATHIIN	COMP
DCNA		2,4-DICHLORO-6-NITRO ANILINE	COMP
DCOC		2,4-DICHLOROBENZOYL CHLORIDE	COMP
DCP	DCP-LA	DIRECT CURRENT ARGON PLASMA LASER ABLATION	INSTR
DCP		DICAPRYLPHTHALATE	COMP
DCP-OES		DIRECT CURRENT ARGON PLASMA OPTICAL EMISSION SPECTROMETRY	METH
DCPA		1,3,5,6-TETRACHLORO-1,4-BENZODICARBOXYACID DIMETHYLESTER	COMP
DCPC	DMC	4,4'-DICHLORO DIPHENYL METHYL CARBINOL	COMP
DCR		DYNAMIC CHARGE RESTORATION	INSTR
DCT		DICHLOROTOLUENE DIVERSE	COMP
DCTA	DCTE	TRANS-1,2-DIAMINO CYCLOHEXANE TETRAACETIC ACID	COMP
DCTBF	2,4-DCTBF	2,4-DICHLOROBENZOTRIFLUORIDE	COMP
DCTC	3,4-DCTC	3,4-DICHLOROBENZOTRICHLORIDE & DIVERSE	COMP
DCTE	DCTA	TRANS-1,2-DIAMINO CYCLOHEXANTETRAESSIGSAURE	COMP
DCTFA		1,3-DICHLORO 1,1,3,3-TETRAFLUOROACETONE	COMP

Abkürzung Akromym abbreviation akronym	Alternative alternative	Bedeutung meaning	Zuordnung related section
DCTL		DIRECT COUPLED TRANSITOR LOGIC	ELECT
DCTP		2-DEOXYCYTIDINE	COMP CLINCHEM
DCU		N,N-DICHLORO URETHANE	COMP
DCUP		DICUMYLPEROIDE	COMP
DDA		4,4'-DICHLORODIPHENYLACETIC ACID	COMP
DDA		DEUTSCHER DAMPFKESSEL AUSSCHUß (D)	ORGANIS
DDAVP		1-DESAMINO-8-D-ARGININE VASOPRESSIN	COMP
DDB		2,3-DIMETHOXY-1,4-BIS(DIMETHYLAMINO)-BUTANE	COMP
DDC		DIETHYLDITHIOCARBAMATE	COMP
DDC		DIRECT DIGITAL CONTROL	INSTR
DDCI		(2-(3,4DICHLOROPHENYL)-2-HYDROXYETHYL) DIMETHYLISOPROP.AMMONIUMIODIDE	COMP
DDD	TDE	DICHLORODIPHENYL DICHLORETHANE (2,4' & 4,4' OR o,p' & p,p')	COMP
DDD		2,2'-DIHYDROXY-6,6'-DINAPHTHYLSULFIDE	COMP
DDE		DICHLORODIPHENYL DICHLORETHENE (2,4' & 4,4')	COMP
DDH		1,3-DIBROMO-5,5-DIMETHYLHYDANTOIN	COMP
DDK		DYNAMISCHE DIFFERENZKALORIMETRIE	METH
DDM	4,4'& 2,4'	BIS (4-CHLOROPHENYL)METHANE	COMP
DDM		DIPHENYLDIAZOMETHANE	COMP
DDMS	2,4'& 4,4'	1-CHLORO-2-(2-CHLOROPHENYL)-2-(4'CHLOROPHENYL)ETHANE	COMP
DDMU	2,4'& 4,4'	1-CHLORO-2-(2-CHLOROPHENYL)-2-(4'-CHLOROPHENYL)-ETHENE	COMP
DDNU		2,2-BIS-(4-CHLOROPHENYL)ETHYLENE	COMP
DDOH		4,4'-DICHLORODIPHENYLETHANOL	COMP
DDP		DI-n-DECYL PHTHALATE	COMP
DDP	CACP	CIS / TRANS DIAMMINE DICHLORO PLATINIUM	COMP
DDQ		2,3-DICHLORO-5,6-DICYANO-1,4-BENZOQUINONE	COMP
DDS		4,4'-DIAMINODIPHENYL SULFONE =DIAPHENYLSULFONE	COMP
DDS		DIHYDROXYDIPHENYL SULFONE	COMP
DDS		1,1-DICHLORODIMETHYLSULFONE	COMP
DDSA		DODECENYLSUCCINIC ANHYDRIDE	COMP
DDT		DICHLORODIPHENYL-1,1,1-TRICHLOROETHANE (2,4' & 4,4')	COMP

Abkürzung Akromym / abbreviation akronym	Alternative / alternative	Bedeutung / meaning	Zuordnung / related section
DDTT		5,6-DIPHENYL-2,3-DIHYDRO-AYM-TRIAZIN-3-THION	COMP REAG
DDVP		DIMETHYLO-(2,2-DICHLOROVINYL) PHOSPHATE (=DICHLOVOS)	COMP
DDW		DESTILLED DEIONIZED WATER	MATER
DDZ		ALPHA,ALPHA-DIMETHYLDIMETHOXYBENZYLOXYCARBONYL	COMP
DE		DIESEL EXHAUST	ENVIR MATER
DE	2-DE	TWO-DIMENSIONAL ELECTROPHORESIS	METH
DEA		N,N-DIETHYLANILINE	COMP
DEAA		N,N-DIETHYLACETOACETAMIDE	COMP
DEAC		DIETHYLALUMINIUMCHLORIDE	COMP
DEAD		DIETHYL AZO DICARBOXYLATE	COMP
DEAE		DIETHYLAMINOETHYL (LIGAND)	COMP
DEAH		DIETHYLALUMINIUM HYDRIDE	COMP
DEAI		DIETHYLALUMINIUM IODIDE	COMP
DEAP		2,2-DIETHOXYACETOPHENONE	COMP
DEASA		N,N-DIETHYLANILINE-3-SULFONIC ACID	COMP
DEB		DISCHARGE ELECTRON BREMSSTRAHLUNG	METH
DEC		DONNAN EXCLUSION CHROMATOGRAPHY	METH
DEC		DIETHYLAMINOETHYL CHLORIDE	COMP
DECHEMA		DEUTSCHE GESELLSCHAT FUER CHEMISCHES APPARATEWESEN (D)	ORGANIS
DEDM		DIETHYL DIAZOMALONATE	COMP
DEE		2,4-DIMETHYLGLUTARSÄURE DIETHYLESTER	COMP
DEFT		DRIVEN EQUILIBRIUM FOURIER TRANSFORM NMR	METH
DEFT		DIRECT EPIFLUORESCENT FILTER TECHNIQUES	METH
DEG		DIETHYLENE GLYCOL	COMP
DEGA		DIETHYLENE GLYCOL ADIPATE	COMP
DEGESCH	DeGeSch	DEUTSCHE GESELLSCHAFT FÜR SCHÄDLINGSBEKÄMPFUNG (D)	ORGANIS
DEGMEE		DIETHYLENE GLYCOL MONOMETHYLESTER	COMP
DEGS		DIETHYLENE GLYCOL SUCCINATE	COMP
DEGSE		DIETHYLENE GLYCOL SEBACATE	COMP
DEGUSSA		DEUTSCHE GOLD UND SILBERWARENSCHEIDE ANSTALT (D)	COMPANY
DEHP	DOP	BIS (2-ETHYLHEXYL) PHTHALATE	COMP

Abkürzung Akromym abbreviation akronym	Alternative alternative	Bedeutung meaning	Zuordnung related section
DEII		DIETHYLINDOLOINDOLE	COMP
DELFIA		DISSOCIATION ENHANCED LANTHANIDE	METH
DEM	DEMIN	DEMINERALIZED WATER	MATER
DEN	DENA	N,N-DIETHYL NITROSAMINE	COMP
DENA	DEN NDEA	N,N-DIETHYL NITROSAMINE	COMP
DENS		DIFFUSE LETTIC NEUTRON SCATTERING	METH
DEP		DIELECTROPHORESE	INSTR
DEP		DIETHYL-PYROCARBONAT	COMP
DEP		DIRECT EXPOSURE PROBE (MS)	METH
DEP		DIETHYL PHTHALATE	COMP
DEP		DEPARTMENT OF ENVIRONMENTAL PROTECTION (US)	INSTI
DEPC		DIETHYL PHOSPHONO CYANIDATE (=DIETHYLPHOSPHORYL CYANIDE)	COMP REAG
DEPHA		DI-(2-ETHYLHEXYL) PHOSPHORIC ACID	COMP
DEPT-TECHN		DISTORTIONLESS ENHANCEMENT OF POLARIZATION TRANSFER TECHN (NMR)	INSTR
DES		DEPARTMENT OF EDUCATION AND SCIENCE (GB)	INSTI
DES		DIETHYLSTILBESTROL	COMP
DES-Na	SES	2,4-DICHLOROPHENOXY ETHYL HYDROGENSULFATE SODIUM SALT	COMP
DESS		BIS-(2-ETHYLHEXYL) SULFOSUCCINATE SODIUM	COMP
DESS		DIETHYL SUCCILYLSUCCINATE	COMP
DESY		DEUTSCHES ELECTRONENSYNCHROTRON HAMBURG (D)	INSTI
DET		DIETHYLTRYPTAMIN	COMP CLINCHEM
DET		DIETHYLTARTRATE	COMP
DETA	m-DETA	N,N-DIETHYL-3-TOLUAMIDE	COMP
DETA		DIETHYLENE TETRAMINE	COMP
DETA		DIELECTRIC THERMAL ANALYSIS	METH
DEV		DEUTSCHE EINHEITSVERFAHREN FUER WASSERUNTERSUCHUNG	NORM
DEXA	DExA	DEUTSCHER AUSSCHUß EXPLOSIONSGESCHÜTZTE ELEKTRISCHE ANLAGEN (D)	ORGANIS
DF		DARK FIELD	INSTR
DFA		DOUBLE FILTER ANALYZER	INSTR

Abkürzung Akromym	Alternative	Bedeutung	Zuordnung
abbreviation akronym	alternative	meaning	related section
DFBO		DEUTSCHE FORSCHUNGSGES. FÜR BLECHBEARBEIT.UND OBERFLÄCHENBEHANDLUNG	ORGANIS
DFDD		1,1-DICHLORO-2,2-BIS(4-FLUOROPHENYL)ETHANE	COMP
DFDT		1,1,1-TRICHLORO-2,2-BIS(4-FLUOROPHENYL)ETHANE	COMP
DFG		DEUTSCHE FORSCHUNGS GEMEINSCHAFT (D)	INSTI
DFHC		DARK FIELD HOLLOW CORE	INSTR
DFN		DEUTSCHES FORSCHUNGSNETZ	ELECT TECHN
DFOM		DEFEROXAMINE	COMP
DFP		DIISOPROPYL FLUORO PHOSPHATE	COMP
DFTPP		DECAFLUORO TRIPHENYL PHOSPHINE	COMP
DFVLR		DEUT.FORSCHUNGS UND VERSUCHSANSTALT FÜR LUFT UND RAUMFAHRT (D)	INSTI
DG	2DG	DEOXYGLUCOSE	COMP
DG		DEUTSCHE GESELLSCHAFT FÜR GALVANOTECHNIK E.V. (D)	ORGANIS
DG		DIRECTION GENERALE / DIRECTORATE GENERAL (EG)	INSTI
DGA		DEUTSCHER DRUCKGAS-AUSSCHUß (D)	ORGANIS
DGAVP		DESGLYZINAMID-ARGININ-VASOPRESSIN	CLINCHEM
DGDG		DIGALACTOSYL DI GLYCERIDE	COMP CLINCHEM
DGE		DEUTSCHE GESELLSCHAFT FÜR ERNÄHRUNG (D)	ORGANIS
DGF		DEUTSCHE GESELLSCHAFT FUER FETTWISSENSCHAFT (D)	NORM ORGANIS
DGFH	DGfH	DEUTSCHE GESELLSCHAFT FÜR HOLZFORSCHUNG	ORGANIS
DGHM		DEUTSCHE GESELLSCHAFT FÜR HYGIENE UND MIKROBIOLOGIE (D)	ORGANIS
DGKCH		DEUTSCHE GESELLSCHAFT FÜR KLINISCHE CHEMIE (D)	ORGANIS
DGMK		DEUTSCHE GESELLSCHAFT FÜR MINERALÖLWISSENSCHAFT UND KOHLECHEMIE (D)	ORGANIS
DGNÄ		DEUTSCHE GESELLSCHAFT NATURFORSCHER UND ÄRZTE (D)	ORGANIS
DGPH		DEUTSCHE GESELLSCHAFT FUER PHYSIK (D)	ORGANIS

Abkürzung Akromym abbreviation akronym	Alternative alternative	Bedeutung meaning	Zuordnung related section
DGQ		DEUTSCHE GESELLSCHAFT FÜR QUALITÄTSFORSCH. PFLANZL.NAHRUNGSMITTEL (D)	ORGANIS
DGRST		DELEGATION GENERAL A LA RECHERCHE SCIENTIFIQUE ET TECHNIQUE (F)	ORGANIS
DGU		DEUTSCHE GESELLSCHAFT FÜR UMWELTSCHUTZ E.V. (DÜSSELDORF,D)	ORGANIS
DGWK		DEUTSCHE GESELLSCHAFT FÜR WARENKENNZEICHNUNG (BERLIN,D)	ORGANIS
DGZFP	DGZfP	DEUTSCHE GESELLSCHAFT FUER ZERSTOERUNGSFREIE PRUEFUNG (D)	ORGANIS
DHA		DI-DEHYDRO ARACHIDON ACIDS (DIFFERENT)	COMP
DHA		DIHYDROXY ACETONE	COMP
DHA		DEHYDRO ACETIC ACID	COMP
DHA		9,10-DEHYDRO ANTHRACENE	COMP
DHAP		DIHYDROXY ACETON PHOSPHAT	COMP
DHAQ		MITOXANTRONE	COMP
DHBA		DIHYDROXY BENZYLAMINE DIVERSE	COMP
DHBP		2,4-DIHYDROXY BENZOPHENONE	COMP
DHE		DIHAEMATO PORPHYRIN ETHER	COMP
DHEBA		N,N'-(1,2-DIHYDROXY ETHYLENE)-BIS-ACRYLAMIDE	COMP
DHET		DIHYDRO ERGOTOXINE	COMP
DHEW		DEPARTMENT OF HEALTH,EDUCATION AND WELFARE (US)	INSTI
DHF		DEHYDRO FOLIC ACID	COMP CLINCHEM
DHFR		DIHYDRO FOLATE REDUCTASE	CLINCHEM
DHHS		DEPARTMENT OF HEALTH AND HUMAN SERVICES (US)	INSTI
DHI		DEUTSCHES HYDROGRAPHISCHES INSTITUT (D)	INSTI
DHIC		DIHYDRO ISOCODEINE	COMP
DHL		DESOXYCHOLATE HYDROGENSULFATE LACTOSE	COMP CLINCHEM
DHN		5,12-DIHYDRO NAPHTHACENE	COMP
DHP		DIHYDRO PTEROINSAEURE	COMP CLINCHEM
DHP		DIHEPTYL PHTHALATE	COMP
DHP		DEHYDRO PYRAN	COMP
DHS		DIHEXYL SEBACATE	COMP
DHSM	DST	DEHYDRO STREPTO MYCIN	COMP

Abkürzung Akromym abbreviation akronym	Alternative alternative	Bedeutung meaning	Zuordnung related section
DHSS		DEPARTMENT OF HELATH AND SOCIAL SECURITY (US)	INSTI
DHT		DEHYDRO TESTOSTERON	COMP CLINCHEM
DIAD		DIISOPROPYL DIAZODICARBOXYLATE	COMP
DIB		1,3-DIPHENYL ISOBENZO FURAN	COMP
DIBA		DIISOBUTYL ADIPATE	COMP
DIBAC		DIISO BUTYL ALUMINIUM CHLORIDE	COMP
DIBAH	DIBAL	DIISOBUTYL ALUMINIUM HYDRIDE	COMP
DIBK		DIISOBUTYL KETON =2,6-DIMETHYL-HEPTAN 4-ONE	COMP
DIBP		DIISOBUTYL PHTHALATE	COMP
DIC		DIFFERENTIAL INTERFACE CONTRAST	INSTR
DIC		DIMETHYLAMINO ISOPROPYL CHLORIDE	COMP
DIC	DTIC	5-(3,3-DIMETHYL-1-TRIAZENYL) 1H-IMIDAZOLE 4-CARBOXAMIDE	COMP
DIDA		DIISODECYL ADIPATE	COMP
DIDP		DIISODECYLPHTHALATE	COMP
DIDT		5,6-DIHYDRO-3H-IMIDAZO(2,1-C) 1,2,4-DITHIAZOLE 3-THIONE	COMP
DIE		DIRECT INJECTION ENTHALPY	THEOR
DIEDA	DADA	DIISOPROPYLAMMONIUM DICHLORO ACETATE	COMP
DIET	DI-ET	N,N-DIETHYL-p-PHENYLENEDIAMINE	COMP
DIGLYME	Diglyme	DIETHYLENE GLYCOL DIMETHYL ETHER	COMP
DIHT		DEUTSCHER INDUSTRIE UND HANDELSTAG (D)	ORGANIS
DIK		DEUTSCHES INSTITUT FUER KUNSTSTOFFTECHNOLOGIE (D)	ORGANIS
DIM-METH		DIATOMICS IN MOLECULES METHODE	THEOR
DIMDI		DEUTSCH.INSTITUT FÜR MEDIZINISCHE DOKUMENTATION UND INFORMATION (D)	INSTI
DIN		DEUTSCHES INSTITUT FUER NORMUNG (D)	ORGANIS NORM
DINS		DIFFUSE INELASTIC NEUTRON SCATTERING SPECTROSCOPY	METH
DINST		DIN-INFORMATIONSSYSTEM TECHNIK (DIN,D)	LIT
DIOA		DI ISO OCTYL ADIPATE	COMP
DIOM		DI ISOOCTYL MALEATE	COMP
DIOP		DI ISOOCTYL PHTHALATE	COMP

Abkürzung Akromym	Alternative	Bedeutung	Zuordnung
abbreviation akronym	alternative	meaning	related section
DIOP		2,3-O-ISOPROPYLIDENEN-2,3-DIHYDROXY 1,4-BIS-(DIPHENYLPHOSPHINO)BUTANE	COMP
DIOS		DI ISOOCTYL SEBAZATE	COMP
DIP		DUAL IN LINE PACKAGE	ELECT
DIP		DIRECT INSERT PROBE	INSTR
DIPA		2,6-DICHLORO INDOPHENYL ACETATE	COMP
DIPA-DCA	DIEDA	DIISOPROPYLAMMONIUMDICHLORO ACETATE	COMP
DIPC		DIMETHYLAMINOISOPROPYL CHLORIDE	COMP
DIPHOS		ETHYLENE BIS(DIPHENYLPHOSPHINE)	COMP
DIPT		DIISOPROPYL TARTRATE	COMP
DIS		DRAFT INTERNATIONAL STANDARD (ISO-DIN,INT)	NORM
DISC-PAGE		DISCONTINUOUS POLYACRYLAMIDE ELECTROPHORESIS	METH
DISEM		DIFFRACTED BEAMS FROM SECONDARY ELECTRONS IN SEM	METH
DISN		DI IMINO SUCCINONITRILE	COMP
DIT		DI IODO TYROSINE	COMP
DITC		1,4-PHENYLENE DIISOCYANATE	COMP
DITR		DEUTSCHES INFORMATIONSZENTRUM FUER TECHNISCHE REGELN (DIN,D)	ORGANIS
DIVEMA		DIVINYLETHER-MALEIC ACID ANHYDRID COPOLYMER	MATER POLYM
DIW		DEIONIZED WATER	MATER
DK		DIELEKTRIZITAETSKONSTANTE	INSTR
DKD		DEUTSCHER KALIBRIER DIENST (D)	ORGANIS
DKE		DEUTSCHE ELEKTROTECHNISCHE KOMMISSION (VDE - D)	ORGANIS
DKFZ		DEUTSCHES KREBSFORSCHUNGSZENTRUM HEIDELBERG (D)	INSTI
DKG		DEUTSCHE KAUTSCHUK GESELLSCHAFT (D)	ORGANIS
DKI		DEUTSCHES KUNSTSTOFF-INSTITUT (D)	ORGANIS
DKI		DEUTSCHE KOMMISSION FÜR INGENIEURAUSBILDUNG (D)	ORGANIS
DKP		5-BENZYL-3,6-DIOXO-2-PIPERAZINE ACETIC ACID (SWEATENER)	COMP
DKSG		O-ACETOACETYLPHENOL-SILICA GEL =DIKETONE SILICA GEL	MATER
DKV		DEUTSCHER KLIMATECHNISCHER VEREIN (D)	ORGANIS

Abkürzung Akromym	Alternative	Bedeutung	Zuordnung
abbreviation akronym	alternative	meaning	related section
DL50	LD50	DOSIS LETAL 50%	CLINCHEM
DLI		DIRECT LIQUID INLET	INSTR
DLI		DEUTSCHES LACKINSTITUT (D)	ORGANIS
DLS		DYNAMIC LIGHT SCATTERING	METH
DLTDP		DILAURYLTHIODIPROPIONATE (ANTIOX)	COMP
DLTMA		DYNAMIC LOAD THERMOMECANICAL ANALYSIS	METH
DLVO		DERJAGUIN & LANDAU,VERWEY & OVERBECK	THEOR
DMA		DIMETHYL ARSONIC ACID	COMP
DMA	VMA	3,4-DIHYDROXY MANDELIC ACID	COMP
DMA		9,10-DIMETHYLANTHRACENE	COMP
DMA		DIMETHYL ANILINE DIV.	COMP
DMA		DIFFERENTIELLER MOBILITÄTS PARTIKEL ANALYSATOR	TECHN
DMA	DMAC	DIMETHYLACETAMIDE	COMP
DMA		DIMETHYLANISOLE OFT 2,6-	COMP
DMA-DEA		N,N-DIMETHYLACETAMIDE-DIETHYL ACETAL	COMP
DMAA		DIMETHYL ALLYL AMINE	COMP
DMAA		N,N-DIMETHYLACETOACETAMIDE	COMP
DMAB		4-DIMETHYLAMINOBENZALDEHYDE	COMP
DMABN		4-(DIMETHYLAMINO) BENZONITRILE	COMP
DMAC	DMA	DIMETHYLACETAMIDE	COMP
DMAD		DIMETHYL ACETYLENEDICARBOXYLATE	COMP
DMAEMA	DMA-EMA	2-DIMETHYLAMINOETHYL METHACRYLATE	COMP
DMAP		4-DIMETHYL AMINO PYRIDINE	COMP
DMAP		DIMETHYLAMINO PROPYLAMINE	COMP
DMAP		4-DIMETHYLAMINO PHENOL	COMP
DMAPN		3-DIMETHYLAMINO PROPIONITRIL	COMP
DMB		(3,3-DIMETHYL BUTYL)DIMETHYLSILOXYL (GC-PHASE)	COMP
DMBA		DIMETHYL-BENZYLYLAMINE	COMP
DMBA		DIMETHYL BENZANTHRACENE	COMP
DMBPPD		N-(1,3-DIMETHYL BUTYL)-N'-PHENYL-p-PHENYLENE DIAMINE	COMP
DMBS		DIMETHYL BENZA THIAZOLYL SULPHENE OXIDE	COMP
DMC		4'4'-DICHLORODIPHENYLMETHYL ARBINOL	COMP
DMC		BETA-BETA-DIMETHYLCYSTEINE =3 MERCAPTO-D-VALINE	COMP

Abkürzung Akromym abbreviation akronym	Alternative alternative	Bedeutung meaning	Zuordnung related section
DMD	2,4' DMD	1-CHLORO-2-(2CHLOROPHENYL)-2-(4'CHLOROPHENYL)-ETHANE	COMP
DMD		DUCHENNE MUSKELDYSTROPHIE	CLINCHEM
DMDS	DTDM	4,4'-DITHIOMORPHOLINE	COMP POLYM
DME		DROPPING MERCURY ELECTRODE	INSTR
DME		1,2-DIMETHOXYETHANE = ETHYLENEGLYCOL DIMETHYLETHER	COMP
DME	4,4'	1-CHLORO-2,2-BIS(4-CHLOROPHENYL)-ETHANE	COMP
DME		2,4-DIMETHYLGLUTARSÄURE DIMETHYLESTER	COMP
DMECS		DIMETHYLETHYLCHLOROSILANE	COMP
DMEU	DMI	N,N'-DIMETHYLETHYLENEUREA = 1.3-DIMETHYL-2-IMIDAZOLIDINONE	COMP
DMF		N,N-DIMETHLYFORMAMIDE	COMP
DMF-DMA		N,N-DIMETHYLFORMAMIDE DIMETHYL ACETAL	COMP
DMHPPD		N,N'-(3-METHYL HEPTYL)-p-PHENYLENE DIAMIN	COMP
DMI	DMEU	N,N'-DIMETHYLETHYLENE UREA (=1,3-DIMETHYL-2-IMIDAZOLIDONE)	COMP
DMM		DIGITAL MULTIMETER	INSTR
DMMP		DIMETHYL METHANOL PHOSPHATE	COMP
DMN	DMNA	DIMETHYLNITROSAMIN	COMP
DMOA		N,N-DIMETHYLOCTYLAMINE	COMP
DMP		DIMETHYL PHENOL DIVERSE	COMP
DMP		DIMETHYLPHTHALATE	COMP
DMP		DIMETHYL PYROCARBONATE	COMP
DMP		2,2-DIMETHOXYPROPANE	COMP
DMP-30		2,4,6-TRIS(DIMETHYLAMINOETHYL)PHENOL	COMP
DMPA		O-2,4-DICHLOROPHENYL-O-METHYLTHIO-PHOSPHORIC ACID-ISOPROPYL AMIDE	COMP
DMPA		2,2-DIMETHOXY-2-PHENYLACETOPHENONE	COMP
DMPC		DIMETHYLAMINOPROPYL CHLORIDE	COMP
DMPE		1,2-BIS (DIMETHYLPHOSPHINO)ETHANE	COMP
DMPO		5,5-DIMETHYL-1-PYRROLL-1-OXIDE = 5,5-DIMETHYL-1-PYROLLINE-N-OXIDE	COMP
DMPO		DIMETHYL PHOSPHATE	COMP
DMPP		1,1-DIMETHYL-4-PHENYLPIPERAZINIUM IODIDE	COMP

Abkürzung Akromym abbreviation akronym	Alternative alternative	Bedeutung meaning	Zuordnung related section
DMPS		2,3-DIMERCAPTO-1-PROPANESULFONIC ACID	COMP
DMPT		N,N-DIMETHYLPROTRIPTYLINE-LIGAND	COMP
DMPU		1.3-DIMETHYLTETRAHYDRO-2(1H)PYRIMIDINE (=N,N'DIMETHYLPROPYLENEUREA)	COMP
DMQ		DIMETHYL AMINO QUINOLINE	COMP
DMS		DEHNUNGS MESS STREIFEN	INSTR
DMS	DMS-	DIMETHYL SILYL-LIGAND	COMP
DMS		DIMETHYLSULFONE	COMP
DMS		4,6-DIMETHOXYBENZENE-1,3-DISULFONYL CHLORIDE	COMP
DMSO		DIMETHYL SULFOXIDE	COMP
DMSS		DIMETHYL SUCCINYLSUCCINATE	COMP
DMT	DMT-	DIMETHOXY TRIPHENYLMETHYL(=TRITYL)-LIGAND	COMP
DMT		DIMETHYL TEREPHTHALATE	COMP
DMT		DYNAMIC MECHANICAL TESTING	METH
DMTA		DYNAMIC MECHANICAL THERMAL ANALYZER	INSTR
DMTAR		4-(4,5-DIMETHYL-2-THIALZOLYL AZO)RESORCINE	COMP
DMTD		2,5-DIMERCAPTO-1,3,4-THIADIAZOLE	COMP
DMTDS		2,5-DIMERCAPTO-1,3,4 THIADIAZOLE DISULFIDE	COMP
DMTR		DIMETHOXYTRITYL-LIGAND	COMP
DMTSF		DIMETHYL(METHYLTHIO)SULFONIUM FLUOROBORATE	COMP
DN		GUTMANN DONOR NUMBER	THEOR
DNA		DIANYL ADIPATE (ACRONYM BEREITS BEKANNT BELEGT)	CLINCHEM
DNA		DEUTSCHER NORMEN AUSSCHUSS (DIN,D)	NORM ORGANIS
DNA		DESOXYRIBONULEIC ACID	COMP
DNAP		2-(2-PENTYL)-4,6-DINITROPHENOL	COMP
DNAP		4-(2',4'-DINITROPHENYLAZO)-9-PHENANTHROL	COMP
DNBC		3,5-DINITRO BENZOYL CHLORIDE	COMP REAG
DNBP		2-(2-BUTYL)-4,6-DINITRO PHENOL	COMP
DNBPG		(R)-N-3,5-DINITROBENZOYL PHENYL GLYCINE	COMP
DNBS		2,4-DINITRO BENZENESULFONIC ACID	COMP
DNBSC		2,4-DINITROBENZENESULFENYL CHLORIDE	COMP
DNCLA		3',4'-DEOXYNANLAUDANASOLINE CARBOXYLIC ACID	COMP CLINCHEM

Abkürzung Akromym	Alternative	Bedeutung	Zuordnung
abbreviation akronym	alternative	meaning	related section
DNF	FDNB DNFB	2,4-DINITROFLUOROBENENE	COMP
DNFA		2,4-DINITRO-5-FLUORO ANILINE	COMP
DNFB	DNF	2,4-DINITRO FLUORO BENZENE	COMP
DNMR		DYNAMIC NUCLEAR MAGNETIC RESONANCE	METH
DNMR	2D-NMR	TWODIMENSIONAL NUCLEAR MAGNETIC RESONANCE SPECTROSCOPY	METH
DNOC		DINITRO-o-CRESOL DIVERSE	COMP
DNOCHP		2-CYCLOHEXYL-4,6-DINITROPHENOL	COMP
DNOCHP		DI-N-OCTYL-N-DECYL PHTHALATE	COMP
DNOCP		DINOCAP = 2-BUTENOIC ACID 2-(1-METHYLHEPTYL) 4,6-DINITROPHENYL	COMP
DNP		DESOXYRIBONUCLEIC ACID PROTEIN COMPLEX	CLINCHEM
DNP		DINONYL PHTHALATE	COMP
DNP	DNP-	DINITROPHENYL-LIGAND	COMP
DNPA		DINITROPROPYLACYLATE	COMP
DNPBA		3,5-DINITROPEROXYBENZOIC ACID	COMP
DNPC	2,6-DNPC	2,6-DINITRO-p-CRESOL	COMP
DNPF	DNF	2,4-DINITROFLUOROBENZENE	COMP
DNPH		2,4-DINITROPHENYLHYDRAZINE	COMP REAG
DNPT		N,N'-DINITROSO PENTAMETHYLENE TETRAMINE	COMP
DNS		5-DIMETHYLAMINO-1-NAPHTHALENE SULFONIC ACID	COMP
DNS		4,4'DINITROSTILBENE-2,2'DISULFONIC ACID	COMP
DNS-BBA		N-DANSYL-3-AMINOBENZENEBORANIC ACID	COMP
DNSA		DANSYLAMIDE (=5-DIMETHYLAMINO-NAPHTHALIN-1 SULFONAMIDE)	COMP
DNSCL	DANS	DANSYLCHLORIDE = 5-DIMETHYLAMINO-NAPHTHALENE-1-SULFOCHLORIDE)	COMP REAG
DNT		DINITROTOLUENE	COMP
DNTC		4-DIMETHYLAMINO-1-NAPHTHYL ISOTHIOCYANATE	COMP REAG
DOA		2-(3',7'-DIMETHYL-2',6'OCTADIENYL)AMINO ETHTANOL	COMP
DOA		DIOCTYL ADIPATE (=DI-(2-ETHYLHEXYL)ADIPATE)	COMP
DOC		DISSOLVED ORGANIC CARBON	MATER
DOC		DIRECT ON COLUMN INJECTOR	INSTR
DOC		DEPARTMENT OF COMMERCE (US)	INSTI

Abkürzung Akromym abbreviation akronym	Alternative alternative	Bedeutung meaning	Zuordnung related section
DOCA		DEOXYCORTICOSTERONE ACETATE	COMP
DOCL		DISSOLVED ORGANIC CHLORINE	MATER
DODPA		4,4'-DIOCTYL DIPHENYL AMINE	COMP
DODS		DIFFERENT ORBITALS FOR DIFFERENT SPINS	THEOR
DOE		DEPARTMENT OF ENERGY (US)	INSTI
DOET		2,5-DIMETHOXY-4-ETHYL PHENYL 2-AMINOPROPANE = 2,5-DIMETHOXY-4-ETHYL-AMPHETAMINE	CLINCHEM COMP
DOF		DI-OCTYL-FUMARATE (=DI-2-ETHYLHEXYL FUMARATE)	COMP
DOIP		DI-(2-ETHYLHEXYL)-ISOPHTHALTE	COMP
DOL		DEPARTMENT OF LABOR (US)	INSTI
DOM		DISSOLVED ORGANIC MATTER	MATER
DOM		DI-2-ETHYLHEXYL MALEAT (=DI-OCTYL MALEATE)	COMP POLYM
DOM	STP	2,5-DIMETHOXY-4-METHYL AMPHETAMINE	CLINCHEM COMP
DOMA		3,4-DIHYDROXYMANDELIC ACID	COMP
DON		6-DIAZO-5-OXO-L-NORLEUCIN	COMP
DOP		DISSOLVED ORGANIC PARTICULATE	MATER
DOP	DEHP	DIOCTYLPHTHALATE	COMP
DOPA		3-(3,4-DIHYDROXY PHENYL)-DL-ALANINE	CLINCHEM COMP
DOPA		DIOXTYL PHENYL PHOSPHORIC ACID	COMP
DOPAC		3,4-DIHYDROXY PHENYLACETIC ACID	COMP
DOPAP		DI-(4-(1,1,3,3 TETRAMETHYL BUTYL)-PHENYL) HYDROGENPHOSPHAT	COMP CLINCHEM
DOPEG		DIHYDROXY PHENYLENE GLYCOL	COMP
DOPET		3,4-DIHYDROXYPHENETYL ALCOHOL	COMP
DOPS		3,4-DIHYDROXYPHENYL SERINE	COMP
DOQ		DOPAMINE QUINONE	COMP
DOS		DI-OCTYL SEBACATE (=DI-2-ETHYLHEXYL SEBACATE)	COMP
DOS		DENSITY OF STATES	THEOR
DOS		DEUTSCHE OFFENLEGUNGSSCHRIFT (D)	REG
DOS		DISC OPERATING SYSTEM	ELECT
DOSCA		DOKUMENTATION ÜBER STÖRFÄLLE IN CHEMIEANLAGEN (INFUCHS-UBA,D)	LIT
DOT		DEPARTMENT OF TRANSPORT (US)	INSTI
DOTA		1,4,7,10-TERAAZACYCLODODECANE N,N',N'',N'''-TERAACETAT LIGAND	COMP

Abkürzung Akromym / abbreviation akronym	Alternative / alternative	Bedeutung / meaning	Zuordnung / related section
DOTG		N,N'-DI-O-TOLYL-GUANIDINE	COMP
DOXYL		4,4-DIMETHYL-3-OXAZOLIDIN-2-YLOXY	COMP
DP		DIFFERENTIAL PULSE	INSTR
DP		DEGREE OF POLYMERIZATION	POLYM
DP	2,4-Dp	2,4-DICHLORO PHENOXYPROPANONIC ACID	COMP
DPA	DPA-	PYRIDINE-2,6-DICARBOXLATE LIGAND	COMP
DPA		DEUTSCHES PATENTAMT (D)	INSTI
DPASV		DIFFERENTIAL PULSE ANODIC STRIPPING VOLTAMMETRY	METH
DPB		1,4-DIPHENYL-1,3-BUTADIENE	COMP
DPBAH		N,N'-DIPHENYL BENZAMIDINE HYDROCHLORIDE	COMP
DPC		DIFFERENTIAL PHASE CONTRAST	INSTR
DPC		DIFFERENTIAL PHOTO CALORIMETRY	METH
DPCSV		DIFFERENTIAL PULS CATHODIC STRIPPING VOLTAMMETRY	METH
DPD		N,N-DIETHYL-4-PHENYLENE DIAMINE	COMP
DPDDA		DIPEROXIDODECANDIACID	COMP
DPDM		DIPHENYL DIAZOMALONATE	COMP
DPE		DIPIVALYL EPHEDRINE	COMP
DPEA		2,4-DICHLORO-6-PHENYL PHENOXYETHYL AMINE	COMP
DPEP		DEOXOPHYLLO ERYTHRO AETIO PORPHYRIN	COMP CLINCHEM
DPG		N,N-DIPHENYL GUANIDINE	COMP REAG
DPG		DEUTSCHE PHYSIKALISCHE GESELLSCHAFT (D)	ORGANIS
DPH		1,6-DIPHENYL-1,3,5-HEXATRIENE	COMP
DPH		DIPHENYL HYDANTOIN	COMP
DPHG	DPhG	DEUTSCHE PHARMAZEUTISCHE GESELLSCHAFT (D)	ORGANIS
DPIBF		DIPHENYL ISOBENZO FURAN	COMP
DPIV		DIFFERENTIAL PULSE INVERSE VOLTAMMETRY	METH
DPM	dpm	DECAYS PER MINUTE	INSTR
DPM-		DIPIVALO METHANATO-LIGAND	COMP
DPN	NAD	DIPHOSPHOPYRIDINENUCEOLTID (VERALTET)	COMP CLINCHEM
DPO	PPO	2,5-DIPHENYLOXAZOLE	COMP
DPP		DIAMINO-PHENANTHIDINE	COMP

Abkürzung Akromym abbreviation akronym	Alternative alternative	Bedeutung meaning	Zuordnung related section
DPP		DIFFERENTIAL PULSE POLAROGRAPHY	METH
DPP	DPP-	DIPHENYL PHOSPHINYL-LIGAND	COMP
DPPA		N,N-DI PENTAFLUOROBENZOYL PENTAFLUORO ANILINE	COMP
DPPA		DIPHENYL PHOSPHORYL AZIDE	COMP
DPPC		DIPALMITOLYLPHOSPHATIDYLCHOLINE	COMP
DPPH	DPPH-	DIAMINO PHENANTHIDIUNIUM-CATION	COMP
DPQD		DIBENZOYL-p-QUINONE DIOXIME	COMP
DPS		DIMETHYL POLY SILOXANE	COMP POLYM
DPS		TRANS-p,p'-DIPHENYLSTILBENE	COMP
DPS		DESTAINER POWER SUPPLY	INSTR
DPT		DIPHTHERIE-PERTUSSIS-TETANUS (POLIO)	CLINCHEM
DPTU		DIPHENYL THIOUREA	COMP
DPV		DIFFERENTIAL PULSE VOLTAMMETRY	METH
DPZ		DEUTSCHES PRIMATEN ZENTRUM (D)	INSTI
DQE		DETECTOR QUANTUM EFFICIENCY	INSTR
DR		DIFFUSE REFLECTION	INSTR
DRAM		DYNAMIC RANDOM ACCESS MEMORY	ELECT
DRAW		DIRECT READ AND WRITE MEMORY	ELECT
DRED		DOUBLE ROCKING ELECTRON DIFFRACTION	METH
DRG		DEUTSCHE RÖNTGEN GESELLSCHAFT (D)	ORGANIS
DRIFT		DIFFUSE REFLECTANCE INFRARED FOURIER TRANSFORM SPECTROSCOPY	METH
DRP	D.R.P.	DEUTSCHES REICHSPATENT (D)	REG
DRS		DIFFUSE REFLECTANCE SPECTROMETRY	METH
DRSK		DEUTSCHE RHEINSCHUTZKOMMISSION (D)	INSTI
DS		DANSYL SARKOSINE	COMP CLINCHEM
DSAH		DISUCCINIMIDYL (N,N'-DIACETYL HOMOCYSTEINE)	COMP
DSC		DIFFERENTIAL SCANNING CALORIMETRY	METH
DSC		4,4'-DICHLORSTILBEN	COMP
DSHO	HODS	DERIVATIVE SPECTROMETRY HIGHER ORDER	METH
DSIMS		DYNAMIC SECONDARY ION MASS SECTROMETRY	METH
DSIP		DIRECT/DELTA SLEEP INDUCING PEPTIDE	CLINCHEM
DSM		DYNAMIC LIGHT SCATTERING MODE	INSTR SPECT
DSM		DEUTSCHE SAMMLUNG VON MIKROORGANISMEN (D)	INSTI
DSMA		DI SODIUM METHANE ARSENATE	COMP

Abkürzung Akromym abbreviation akronym	Alternative alternative	Bedeutung meaning	Zuordnung related section
DSP		DITHIOBIS (SUCCINIMIDYL PROPIONATE)	COMP
DSS		4,4-DIMETHYL-4-SILAPENTANE-5-SULFATE =3(TRIMETHYLSILYL)1-PROPANESULFONIC ACID	COMP
DSS		DISUCCINIMIDYL SUBERATE	COMP
DSSG		DIAZONIUM SALT SILICA GEL	MATER
DST		DISUCCINIMIDYL TARTRATE	COMP
DST	DHBM	DIISOPROPYL AMMONIUM DICHLORO ACETATE	COMP
DT		DIPHTHERIE TETANUS	CLINCHEM
DT		DIPHTERIE TOXINE	CLINCHEM
DT	2,4-DT	2-(2,4-DICHLOROPHENOXY) PROPIONIC ACID	COMP
DTA		DIFFERENTIAL THERMAL ANALYSIS	METH
DTA		DEUTSCHE TECHNISCHE AKADEMIE (HILDESHEIM,D)	INSTI
DTAF		4-(4,6-DICHLOR-S-TRIAZIN-2-YL-AMINO) FLUORESCEINE	COMP
DTBP		DI-TERT-BUTYL PEROXIDE = BIS(1,1-DIMETHYLETHYL) PEROXIDE	COMP
DTDD		N,N'-DIHEPTYL-N,N'-6,6'TERTRAMETHYL-4,8-DIOXA-UNDECANE DIAMINE	COMP
DTDM	DMDS	4,4-DITHIO DIMORPHOLINE	COMP
DTE		DITHIOERYTHRIOL	COMP
DTG		DIFFERENTIAL THERMO GRAVIMETRY	METH
DTGS		DEUTERATED TRI GLYCINE SULFATE (IR DETECTION)	INSTR
DTI		DEPARTMENT OF TRADE AND INDUSTRY (GB)	INSTI
DTIC	DIC	5-(3,3-DIMETHYL-1-TRIAZENYL)-1H-IMIDAZOLE-4-CARBOXAMIDE	COMP
DTL		DIODE TRANSISTOR LOGIC	ELECT
DTLC	2DTLC	TWO DIMENSIONAL THIN LAYER CHROMATOGRAPHY	METH
DTMC		4,4'DICHLORO-ALPHA-(TRICHLOROMETHYL)-BENZHYDROL	COMP
DTNB		5,5'-DITHIO-BIS-(2-NITROBENZOIC ACID)	COMP REAG
DTPA		DIETHYLENE TRIAMINE PENTA ACETIC ACID	COMP REAG
DTPT	TPD	DITHIOPROPYL THIAMINE	COMP
DTSCH.PAT	Dtsch.Pat.	PATENT DER DDR	REG
DTSG		DITHIZONE SILICA GEL	MATER

Abkürzung Akromym abbreviation akronym	Alternative alternative	Bedeutung meaning	Zuordnung related section
DTSP		3,3'DITHIO-BIS-(SUCCINIMIDYL PROPIONIC ACID)	COMP
DTT		DITHIOTHREITOL =CLELAND'S REAG.	COMP
DUA		DEUTSCHE UMWELTAKTION E.V. (D)	ORGANIS
DVB		DIVINYLBENZENE	COMP
DVCV		DEUTSCHE VEREINIGUNG FUER CHEMIE UND VERFAHRENSTECHNIK (VDI,D)	ORGANIS
DVG		DEUTSCHE VETERINÄRMEDIZINISCHE GESELLSCHAFT (D)	ORGANIS
DVGW		DEUTSCHER VEREIN DES GAS- UND WASSERFACHES (D)	ORGANIS
DVM		DIGITAL VOLT METER	INSTR
DVM		DEUTSCHER VERBAND FÜR MATERIALPRÜFUNG (D)	ORGANIS
DVS		DEUTSCHER VERBAND FÜR SCHWEIßTECHNIK (D)	ORGANIS
DVT		DEUTSCHER VERBAND TECHNISCH-WISSENSCHAFTLICHER VEREINE (D)	ORGANIS
DVWK		DEUTSCHER VERBAND FÜR WASSERWIRTSCHAFT UND KULTURBAU (D)	ORGANIS
DVWW		DEUTSCHER VERBAND FÜR WASSERWIRTSCHAFT (D)	ORGANIS
DWD		DEUTSCHER WETTERDIENST (D)	INSTI
DWK		DEUTSCHE GESELLS.FÜR WIEDERAUFARBEITUNG VON KERNBRENNSTOFFEN mbH (D)	COMPANY
DWR		DRUCKWASSER REAKTOR	TECHN
DWZ		DEUTSCHES WARENZEICHEN (D)	REG
DX-C		DETOXIN COMPLEX	CLINCHEM
DXE		DIXYLYL ETHANE =1,1-BIS(3,4-DIMETHYLPHENYL)ETHANE	COMP
DYCMOS		DYNAMIC CONTEMTARY METAL OXIDE SEMICONDUCTOR	ELECT
DZW		DEUTSCHE DOKUMENTATIONSZENTRALE WASSER (D)	LIT

Abkürzung Akromym / abbreviation akronym	Alternative / alternative	Bedeutung / meaning	Zuordnung / related section
EA		ELECTRON ASSOZIATION (MASS SPECT)	METH
EA		ENZYME ACTIVITY	CLINCHEM
EA		EARLY ANTIGEN	CLINCHEM
EA-MS		ELECTRON ATTACHEMENT (ASSOZIATION) MASS SPECTROSCOPY	METH
EAA		ETHYL ACETOACETATE	COMP
EAA		N-ETHYLANTHRANILIC ACID	COMP
EAB		EUROPAEISCHES ARZNEIBUCH (EUR)	NORM
EACR		EUROPEAN ASSOCIATION FOR CANCER RESEARCH (NOTTINGHAM)	ORGANIS
EACS		EPSILON-AMINO CAPRONIC ACID	COMP
EADC		ETHYL ALUMINIUM DICHLORIDE	COMP
EAF		ENTERO TOXEGENIC ADHESIVE FACTOR	CLINCHEM
EAG		EXPOSURE ASSESMENT GROUP (EPA,US)	INSTI
EAK		ETHYL AMYL KETONE	COMP
EAMG		EXPERIMENTAL AUTOIMMUNO MYASTHESSIA GRAVIS	CLINCHEM
EAN		EUROPEAN ARTICLE CODE NUMBER	MATER TECHN
EAP		ERYTHROCYTIDE ACID PHOSPHATASE	CLINCHEM
EAPROM		ELECTRICALLY ALTERNABLE READ ONLY MEMORY	ELECT
EARN		EUROPEAN ACADEMIC AND RESEARCH NETWORK (EUR)	ORGANIS
EAS		EXPERIMENT ANALYSIS SYSTEM	ELECT
EASC		ETHYLALUMINIUM SESQUICHLORIDE	COMP
EAU		EIDGENÖSSISCHES AMT FÜR UMWELTSCHUTZ (CH)	INSTI
EAWAG		EIDGEN.ANSTALT FÜR WASSERVERS.,ABWASSERREINIG.& GEWAESSERSCHUTZ (CH)	INSTI
EB		ETHIDIUM BROMID	COMP POLYM
EBA		N-ETHYL-N-BENZYLANILINE	COMP
EBASA		N-ETHYL-N-BENZYLANILINE-4-SULFONIC ACID	COMP
EBC		EUROPEAN BREWERY CONVENTION (EUR)	REG
EBCB		EUROPEAN BANK OF COMPUTERPROGRAMMES IN BIOTECHNOLOGY	CLINCHEM
EBCDIC		EXTENDED BINARY CODED DECIMAL INTERCHANGE CODE	ELECT

Abkürzung Akromym abbreviation akronym	Alternative alternative	Bedeutung meaning	Zuordnung related section
EBIC		ELECTRON BEAM INDUCED CURRENT	INSTR
EBICON		ELECTRON BOMBARDMENT INDUCED CONDUCTIVITY	METH
EBNV	EBNA	EPPSTEIN BARR NUCLEUS VIRUS/ANTIGEN	CLINCHEM
EBSA		p-ETHYLBENZENESULFONIC ACID	COMP
EBV		EPPSTEIN BARR VIRUS	CLINCHEM
EC		EDGEWORTH CRAMER SERIES (CHROMATOG)	THEOR
EC	EG ,EWG	EUROPEAN COMMUNITIES (EUR)	INSTI
EC		ELECTRON CAPTURE	METH
EC		ETHYLCELLULOSE	MATER
EC		ETHYLCARBAMATE	COMP
EC	E.c.	ESCHERICHIA COLI	CLINCHEM
ECA		ENVIRONMENTAL CONTAMNINATIONS ACT (CDN)	REG
ECAMA		EUROPEAN CITRIC ACID MANUFACTURERS ASSOCIATION (BRÜSSEL,CEFIC)	ORGANIS
ECB		ENVIRONMENT COORDINATION BOARD (ACC,UN)	INSTI
ECBO		EUROPEAN CELL BIOLOGY ORGANIZATION (EUR)	ORGANIS
ECC		EUROPEAN COMMUNITY COUNCIL (EG)	INSTI
ECCA		EUROPEAN COIL COATING ASSOCIATION (BRÜSSEL,EUR)	ORGANIS
ECCLS		EUROPEAN COMMITTEE ON CLINICAL LABORATORY STANDARDS (EUR)	ORGANIS NORM
ECCM		ELECTRONIC COUNTER COUNTER MEASURES	ELECT
ECD		ELECTRON CAPTURE DEDECTOR	INSTR
ECD		EXCHANGE CAPACITY PER DRY GRAM	MATER
ECDIN		ENVIRONMENTAL CHEMICALS DATA AND INFORMATION NETWORK (EG)	LIT
ECDIS		EUROPEAN ENVIRONMENTAL CHEMICAL DATA AND INFORMATIONS SYSTEM (EG)	LIT
ECE		ECONOMIC COMMISSION FOR EUROPE (GENEVE,UN)	INSTI
ECE		ELECTROCHEMICAL CHROMATOGRAPHY	METH
ECEA		N-ETHYL-N-CHLOROETHYLANILINE	COMP
ECETOC		EUROPEAN CHEMICAL INDUSTRY ECOLOGY AND TOXICOLOGY CENTER	INSTI
ECF		EXTRA CELLULARE FLUID	CLINCHEM

Abkürzung Akromym abbreviation akronym	Alternative alternative	Bedeutung meaning	Zuordnung related section
ECF		ENZYME CONVERSION FACTOR	CLINCHEM
ECG		ELECTRO CARDIO GRAM	CLINCHEM
ECL		EMITTER COUPLED LOGIC	ELECT
ECL		ELECTRO LUMINESCENCE	INSTR
ECL		EQUIVALENT CHAIN LENGTH (CHROMATOG)	THEOR
ECM		EXTRACELLULAR MATRIX	CLINCHEM
ECMA		EUROPEAN COMPUTER MANUFACTURES ASSOCIATION (EUR)	ORGANIS
ECMA		EUROPEAN CATALYSTS MANUFACTURERS ASSOCIATION (CEFIC)	ORGANIS
ECN		EQUIVALENT CHAIN NUMBER	THEOR
ECNSS	ECNS	POLY-(ETHYLENE SUCCINATE METHYL CYANO-ETHYL-SILOXANE)	MATER
ECOIN		EUROPEAN CORE INVENTORY	TECHN
ECOMA		EUROPEAN COMPUTER MEASUREMENT ASSOCIATION (EUR)	ORGANIS
ECOSS		EUROPEAN CONFERENCE ON SURFACE SCIENCE (EUR)	CONF
ECQAEC		ELECTRONIC COMPOUNDS QUALITY ASSURANCE COMMITTEE	ORGANIS
ECR		ELECTRON CHANNELING PATTERN	METH
ECTFE		ETHYLENE CHLORO TRIFLUORO ETHYLENE	COMP POLYM
ECTL		EMITTER COUPLED TRANSISTOR LOGIC	ELECT
ECU		EUROPEAN CURRENCY UNIT (EC)	TRADE
ECV		EXCHANGE CAPACITY PER UNIT VOLUME	EXPER
ED		ENERGY DISPERSION	INSTR
ED		ELECTRON DIFFRACTION	INSTR
ED50	CD50	EFFEKTIVE DOSIS 50%	CLINCHEM
EDA		ETHYLENE DIAMINE & ETHYLENE DIAMMONIUM-LIGAND	COMP
EDA		EXPLORATIVE DATA ANALYSIS	ELECT EVAL
EDA		ENERGY DISPERSIVE ANALYZER	INSTR
EDAC		1-ETHYL-3(3-DIMETHYLAMINOPROPYL) CARBODIIMIDE	COMP
EDANS		2-AMINOETHYLAMINO-1-NAPHTHALENE SULFONIC ACID DIVERSE	COMP
EDB		ETHYLENE DIBROMIDE	COMP
EDC		ENERGY DISTRIBUTION CURVES	THEOR
EDC		ETHYLENE DICHLORIDE = 1,2-DICHLOROETHANE	COMP

Abkürzung Akromym abbreviation akronym	Alternative alternative	Bedeutung meaning	Zuordnung related section
EDCI		1-ETHYL-3-(3-(DIMETHYLAMINO)PROPYL)-CARBODIIMIDE	COMP
EDDP		O-ETHYL-S,S-DIPHENYL DITHIOPHOSPHATE	COMP
EDDT		ETHYLENE DIAMINE-D-TARTRATE	COMP
EDF		ELECTROPHORESIS DUPLICATING FILM	MATER
EDF		ENVIRONMENTAL DEFENCE FUND (US)	ENVIR
EDI		EIDGENÖSSISCHES DEPARTMENT DES INNEREN	INSTI
EDIT		EUROPEAN DISCUSSIONGROUP OF IHALATIONSSTOXICOLOGISTS (EUR)	ORGANIS
EDL		ELECTRODELESS DISCHARGE LAMP	INSTR
EDM		ELEMENTARANALYSE-DICHTE-MOLMASSE	COMP
EDMA		EUROPEAN DIAGNOSTIC MANUFACTURERS ASSOCIATION (EUR)	ORGANIS
EDMA		ETHYLENEGLYCOL-DIMETHYL ACRYLATE	COMP
EDMZ		EIDGENÖSISCHE DRUCKSACHEN UND MATERIALZENTRALE (CH)	INSTI
EDP	EDV	ELECTRONIC DATA PROCESSING	ELECT
EDP		ELECTROPHORESIS DUPLICATING PAPER	MATER
EDPA		2-ETHYL-3,3-DIPHENYL-2-PROPENYL AMINE	COMP
EDRAW		ERASEABLE DIRECT READ AFTER WRITE MEMORY	ELECT
EDS		ENERGY DISPERSIVE SYSTEM	INSTR
EDTA		ETHYLENE DIAMINO TETRA ACETIC ACID	COMP REAG
EDTN		1-ETHOXY-4-(DICHLORO-s-TRIAZINYL)-NAPHTHALENE	COMP
EDTP		ETHYLENE DIAMINE TETRAPROPANOL	COMP
EDV	EDP	ELEKTRONISCHE DATENVERARBEITUNG	ELECT INSTR
EDV		ELECTRODYNAMIC VENTURI	INSTR
EDX	EDXS	ENERGY DISPERSIVE X-RAY SPECTROSCOPY	METH
EDXD		ENERGY DISPERSIVE X-RAY DIFFRACTION	METH
EDXRA		ENERGY DISPERSIVE X-RAY ANALYSIS	METH
EDXRF		ENERGY DISPERSIVE X-RAY FLUORESCENCE	METH
EDXS	EDX	ENERGY DISPERSIVE X-RAY SPECTROSCOPY	METH
EEB		EUROPEAN ENVIRONMENTAL BUREAU (BRÜSSEL)	INSTI
EEC	EG , EWG	EUROPEAN ECONOMIC COMMUNITY (EUR)	INSTI
EEC	E.C.	ESCHERICHIA COLI	CLINCHEM
EECL		EMITTER-TO-EMITTER COUPLED LOGIC	ELECT

Abkürzung Akromym	Alternative	Bedeutung	Zuordnung
abbreviation akronym	alternative	meaning	related section
EEDQ		1-ETHOXYCARBONYL-ETHOXY-DIHYDROXQUINOLINE	COMP REAG
EEG		ELECTRO ENCEPHALOGRAMM	CLINCHEM
EEHP		2-ETHYL-4-ETHOXY-6-HYDROXY-PYRIMIDIN	COMP
EEK		EIDGENÖSSISCHE ERNÄHRUNGSKOMMISSION (CH)	INSTI
EELS		ELECTRON ENERGY LOST SPECTROSCOPY	METH
EEM		EMISSION EXCITATION MATRIX	INSTR
EFA		ESSENTIAL FATTY ACID	CLINCHEM
EFAM		EIDGENÖSS. FORSCHUNGSANSTALT F. MILCHWIRTSCHAFT (CH)	INSTI
EFB		EUROPEAN FEDERATION OF BIOTECHNOLOGY (EUR)	ORGANIS
EFC		ELECTRONIC FLUID CONTROL	TECHN
EFCE		EUROPEAN FEDERATION OF CHEMICAL ENGINEERING (EUR)	ORGANIS
EFCTC		EUROPEAN FLUOROCARBON TECHNICAL COMMITTEE	ORGANIS
EFEMA		EUROPEAN FOOD EMULSIFIER MANUFACTURERS ASSOCIATION (CEFIC)	ORGANIS
EFG		ELECTRIC FIELD GRADIENT	INSTR
EFLA		EUROPEAN FOOD LAW ASSOCIATION (EUR)	ORGANIS
EFPIA		EUROPEAN FEDERATION OF PHARMACEUTICAL INDUSTRIES' ASSOCIATIONS (EUR)	ORGANIS
EFTA		EUROPEAN FREE TRADE ASSOCIATION (EUR)	INSTI
EG	EC ,EWG	EUROPAEISCHE GEMEINSCHAFT (EUR)	INSTI
EGA		EVOLVED GAS ANALYSIS	METH
EGA		ETHYLENE GLYCOL ADIPATE	COMP
EGA		EIDGENÖSSISCHES GESUNDHEITSAMT (CH)	INSTI
EGF		EPIDERMAL GROWTH FACTOR	CLINCHEM
EGIP		ETHYLENE GLYCOL ISOPHTHALATE	COMP
EGS		ETHYLENEGLYCOL SUCCINATE	COMP
EGSE		ETHYLENE GLYCOL SEBACATE	COMP
EGSP		ETHYLENE GLYCOL SUCCINATE METHTYL-PHENYL SILOXANE	COMP
EGSS		ETHYLENE GLYCOL BIS (SUCCINIMIDYL SUCCINATE)	COMP
EGT		ENTWICKLUNGSGEMEINSCHAFT TIEFLAGERUNG (KFK UND GSF,D)	INSTI

Abkürzung Akromym	Alternative	Bedeutung	Zuordnung
abbreviation akronym	alternative	meaning	related section
EGTA		ETHYLENEGLYCOL BIS(2-AMINOETHYLETHER) N,N,N'N' TETRA ACETIC ACID	COMP
EGTCP		ETHYLENE GLYCOL-TETRACHLORO PHTHALATE	COMP
EH		EPOXYHYDRATASE	CLINCHEM
EHDIS		ELECTROHYDRO DINOMINAL ION SOURCES	INSTR
EHDP		ETHANE-1-HYDROXY-1,1-DIPHOSPHORIC ACID	COMP
EHMO		ENHANCED HUECKEL MOLECULAR ORBITAL	THEOR
EHPRG		EUROPEAN HIGH PRESSURE RESEARCH GROUP (EUR)	ORGANIS
EHRD		ENVIRONMENTAL HEALTH RESEARCH DEPARTMENT (US)	INSTI
EHT		EXTENDED HUECKEL THEORY	THEOR
EHTCP		BIS-(2-ETHYLHEXYL)TETRACHLORO PHTHALATE	COMP
EI		ELECTRON IMPACT	EXPER
EIA		ENZYME IMMUNO ASSAY	INSTR METH
EIC	IEC	ELECTROSTATIC INTERACTION CHROMATOGRAPHY	METH
EICP		EXTRACTED ION CURRENT PROFILE (MASS SPECT)	SPECT
EID		ELECTRON IMPACT DESORPTION	METH
EID		EGG INFECTIONS DOSE	CLINCHEM
EIEC		ENTERO INVASIVE ESCHERICHIA COLI	CLINCHEM
EIMS		ELECTRON IMPACT MASS SPECTROMETRY	METH
EINECS		EUROPEAN INVENTORY OF EXCISTING COMMERCIAL CHEMICAL SUBSTANCES	
EIR		EIDGENÖSSISCHES INSTITUT FÜR REAKTORFORSCHUNG (CH)	INSTI
EIRMA		EUROPEAN INDUSTRIAL RESEARCH MANGEMENT ASSOCIATION (EUR)	ORGANIS
EIS		ELECTRON IMPACT SPECTROSCOPY	METH
EKG	ECG	ELECTRO KARDIO GRAMM	CLINCHEM
EKL		EIDGENÖSSISCHE KOMMISSION FÜR LUFT HYGIENE (CH)	INSTI
EL		ELECTRO LUMINESCENCE	METH
ELA		ELECTRO ACOUSTIC	INSTR
ELAS		ELECTRO ACOUSTIC SPECTROMETRY	METH
ELCD		ELECTRO CONDUCTIVITY DETECTOR	INSTR

Abkürzung Akromym abbreviation akronym	Alternative alternative	Bedeutung meaning	Zuordnung related section
ELCD		ELECTROCHEMICAL DETECTION	INSTR
ELCO		ELECTROLYTE CONDENSATOR	ELECT
ELDOR		ELECTRON ELECTRON DOUBLE RESONANCE	METH
ELEED		ELASTIC LOW ENERGY ELECTRON DIFFRACTION	METH
ELI		ENVIRONMENTAL LAW INSTITUTE (US)	INSTI
ELISA		ENZYME LINKED IMMUNO SORBENT ASSAY	INSTR METH
ELLA		ENZYME LINKED LUMINESCENE ASSAY	METH
ELS		ENERGY LOSS SPECTROSCOPY	METH
ELS	LDV	ELECTROPHONETIC LIGHT SCATTERING	INSTR
ELSA		EUROPEAN LEAD STABILISER MANUFACTURERS ASSOCIATION (CEFIC)	ORGANIS
EM		ERNST MERCK	COMPANY
EM		ELECTRON MICROSCOPY	METH
EMA		ETHYLENE MALEIC ACID ANHYDRIDE	COMP
EMA		ELECTRON MICROPROBE ANALYSIS	METH
EMB		EOSIN-METHYLENBLAU-LACTOSE-SACCHAROSE	CLINCHEM
EMB		2,2'-(1,2-ETHANE DIYLDIIMINO)BIS-1-BUTANOL = ETHAMBUTOL	COMP
EMBL		EUROPEAN MOLECULAR BIOLOGY LABORATORY (EUR)	INSTI
EMBO		EUROPEAN MOLECULAR BIOLOGY ORGANIZATION (EUR)	ORGANIS
EMCV		ENCEPHALOMYOCARDITIS VIRUS	CLINCHEM
EMD		EINZELMAXIMAL DOSIS	CLINCHEM
EMF	EMK	ELECTROMOTORIC FORCE	EANAL THEOR
EMFC		ELECTROMOTORIC FORCE COMPENSATION	INSTR
EMI		ELECTROMAGNETIC INTERFERENCE	INSTR
EMIC		ENVIRONMENTAL MUTAGEN INFORMATION CENTER (US)	INSTI LIT
EMIT		ENZYME MULTIPLIED IMMUNO ASSAY	METH
EMK	EMF	ELEKTROMOTORISCHE KRAFT	EVAL THEOR
EMM		ELECTRON MIRROR MICROSCOPY	METH
EMMA		ELECTRON MICROSCOP MICRO ANALYZER	INSTR
EMP		ELECTRO MAGNETIC PULSE	INSTR
EMP		ELECTRON MICRO-PROBE (ANALYSIS)	METH
EMPA		EIDGEN.MATERIALPRÜF.& VERS.ANSTALT F.INDUS ,BAUWES.& GEWERBE (CH)	INSTI

Abkürzung Akromym	Alternative	Bedeutung	Zuordnung
abbreviation akronym	alternative	meaning	related section
EMPA	ESMA	ELECTRON MICROPROBE ANALYSIS	METH
EMQ		6-ETHOXY-1,2-DIHYDRO-2,2,4-TRIMETHYL QUINOLINE	COMP
EMR		ELECTRO MAGNETIC RADIATION	INSTR
EMR		ENZYME MEMBRANE REACTOR	INSTR
EMS		ETHYLENE METHAN SULFONAL	COMP
EMS		ELECTROMAGNETIC SUSCEPTIBILITY	INSTR
EMUF		EINPLATINEN MICROCOMPUTER F.UNIVERSELLE FESTPROGRAMMANWENDUNG	ELECT
ENA		EUROPEAN NITRATORS ASSOCIATION (CEFIC)	ORGANIS
ENAA		EPITHERMAL NEUTRON ACTIVATION ANALYSIS	METH
ENDOR		ELECTRON NUCLEOUS DOUBLE RESONANCE	METH
ENDS		ENVIRONMENTAL DATA SERVICES (GB)	LIT
ENEA		EUROPEAN NUCLEAR ENERGY AGENCY (EUR)	INSTI
ENFET		ENZYME SENSITIVE FIELD EFFECT TRANSSISTOR	INSTR
ENPI		ENTE NATIONALE PREVENZIONE INFORTUNI (ROM,I)	INSTI
ENSCS		ECOLE NATIONALE SUPERIEURE DE CHEMIE DU STRASBOURG (F)	INSTI
ENU		N-ETHYL-N-NITROSO UREA	COMP
EOCL	EOCl	EXTRACTABLE ORGANIC CHLORINE	COMP
EOF		END OF FILE	ELECT
EOM		EQUATIONS OF MOTION	THEOR
EOM		EXTRACELLULARE ORGANIC MATTER	CLINCHEM
EOQC		EUROPEAN ORGANIZATION FOR QUALITY CONTROL (EUR)	ORGANIS
EOR		ENHANCED OIL RECOVERY	TECHN
EOS		END OF STRING	ELECT
EOX		EXTRACTABLE ORGANIC HALOGEN	MATER
EP		ELECTROPHOSPHORECENCE	INSTR
EPA		EUROPEAN PHOTOCHEMISTRY ASSOCIATION (EUR)	ORGANIS
EPA		EUROPAEISCHES PATENTAMT (MUENCHEN)	INSTI
EPA		ENVIRONMENTAL PROTECTION AGENCY (US)	INSTI
EPA		ETHYLPHENACEMIDE	COMP
EPC		EGG PHOSPHATIDYL CHOLIN	CLINCHEM

Abkürzung Akromym	Alternative	Bedeutung	Zuordnung
abbreviation akronym	alternative	meaning	related section
EPCA		EUROPEAN PETROCHEMICAL ASSOCIATION (CEFIC)	ORGANIS
EPD		ETHYLENE PROPYLENE COPLYMER MIXTURE	MATER POLYM
EPDM		ETHYLENE PROPYLENE DIENE MONOMER POLYMER	MATER
EPEC		ENTEROPATHOGENIC ESCHERICHIA COLI	CLINCHEM
EPF		EUROPEAN POLYMER FEDERATION (EUR)	ORGANIS
EPFL		ECOLE POLYTECHNIQUE FEDERAL LAUSANNE (CH)	ORGANIS
EPI		EUROPEAN PETROCHEMICAL INDUSTRY (CEFIC)	ORGANIS
EPM		ELECTRON PROBEMICROSCOPY	METH
EPMA		ELECTRON PROBE MICRO ANALYSIS	METH
EPN		O-ETHYL-O-(p-NITROPHENYL)THIOBENZENE PHOSPHATE	COMP
EPO		ERYTHROPOIETIN HORMON	CLINCHEM
EPPO		EUROPEAN AND MEDITARANEAN PLANT PROTCTION ORGANIZATION (EUR)	INSTI
EPPS		4-(2-HYDROXYETHYL)-1-PIPERAZINE-PROPANE SULFONIC ACID	COMP
EPR		ELECTRON PROTON RESONANCE	METH
EPROM		ERASEABLE PROGRAMMABLE READ ONLY MEMORY	ELECT
EPS		EUROPEAN PHYSICAL SOCIETY (EUR)	ORGANIS
EPS		ELECTROPHORESIS POWER SUPPLY	INSTR
EPS		EXPANSIBLE POLYSTYRENE	MATER
EPTC		ETHYL-N,N-DIPROPYL THIOCARBAMATE	COMP
EPTD		DIETHYL DIPHENYL THIURAM DISULFIDE	COMP
EPTM		DIETHYL DIPHENYL THIURAM MONOSULFIDE	COMP
EPU		EUROPAEISCHE PATENT UEBEREINKUNFT (EUR)	REG
EPXMA		ELECTRONPROBE X-RAY MICROANALYSIS	METH
ER		ENDOPLASMATIC RETICULUM	CLINCHEM
ER		ENOAT REDUCTASE	CLINCHEM
ERBE		EUROPEAN RADIATION BUDGET EXPERIMENT (AIR)	TECHN ENVIR
ERC		EPOXY RESINS COMMITTEE (APME,EUR)	ORGANIS
ERDA		ENERGY RESEARCH AND DEVELOPMENT ADMINISTRATION (US)	INSTI

Abkürzung Akromym abbreviation akronym	Alternative alternative	Bedeutung meaning	Zuordnung related section
ERDB		ENVIRONMENTAL RELEASE DATA BASE	LIT
ERG.B.	Erg.B.	ERGAENZUNGSBUCH ZUM DEUTSCHEN ARZNEIBUCH (D)	NORM
ERMA		EXTENDED RED MULTI ALKALI	MATER
EROM		ERASABLE READ ONLY MEMORY	ELECT
ERS		EUROPEAN REMOTE SATELLITE (EUR)	TECHN
ERS		EXTERNE REFLEXIONSTECHNIK	INSTR
ERS		EXTERNAL REFLECTION SPECTROMETRY	METH
ERTS		EARTH RESOURCES TECHNOLOGY SATELLITE (EUR)	TECHN
ES	EI	ELECTRONENSTOSS	INSTR
ES		EMISSION SPECTROSCOPY	METH
ESA		EUROPEAN SPACE AGENCY (EUR)	INSTI
ESA		ELECTROSTATIC ANALYZER	INSTR
ESA		ELECTRIC SECTOR ANALYZER	INSTR
ESB	UPB	ENVIRONMENTAL SPECIMEN BANK	ENVIR
ESCA	XPS	ELECTRON SPECTROSCOPY FOR CHEMICAL ANALYSIS	METH
ESCIS		EXPERTENKOMMISSION FUER SICHERHEIT IN DER CHEMISCHEN INDUSTRIE (CH)	INSTI
ESD		EFFECTIVE STANDARD DEVIATION	METH EVAL
ESD		ELECTRON STIMULATED DESORPTION	INSTR
ESDI		ELECTRON STIMULATED DESORPTION OF IONS	METH
ESDIAD		ELECTRON STIMULATED DESORPTION ION ANGULAR DISTRIBUTION	METH
ESDN		ELECTRON STIMULATED DESORPTION OF NEUTRALS	METH
ESEM		ELECTRONSCATTERING ELECTRON MICROSCOPY	METH
ESF		EUROPEAN SCIENCE FOUNDATION (EC)	INSTI
ESFI		EPITAXIAL SILICON FILM ON ISOLATORS	ELECT MATER
ESI		ELECTRON SPECTROSCOPY IMAGING	METH
ESID		ELECTRON STIMULATED ION DESORPTION	METH
ESIE		ELECTRON STIMULATED ION EMISSION	METH
ESMA	EMPA	ELEKTRONENSTRAHL MICROANALYSE	METH
ESMA		EUROPEAN STERATES MANUFACTURERS ASSOCIATION (CEFIC)	ORGANIS
ESOC		EUROPEAN SPACE OPERATION CENTER (EUR)	INSTI

Abkürzung Akromym abbreviation akronym	Alternative alternative	Bedeutung meaning	Zuordnung related section
ESR		ELECTRON SPIN RESONANCE	METH
ESR		ERYTHROCYTE SEDIMENTATION RATE	CLINCHEM
ESRF		EUROPEAN SYNCHROTRON RADIATION FISSION	TECHN
ESRO		EUROPEAN SPACE RESEARCH ORGANIZATION	INSTI
ESSA		ENVIRONMENTAL SCIENCE SERVICES ADMINISTRATION (EUR)	INSTI
ET		ELECTRON TRANSFER	THEOR
ET	ETA	ELECTROTHERMAL ATOMIZATION	INSTR
ETA	ET	ELECTROTHERMAL ANALYSIS	METH
ETAA	ET-AAS	ELECTROTHERMAL ATOMIC ABSORPTION	METH
ETAD		ECOLOG.& TOXICOL. ASSOC.OF THE DYSTUFFS MANUFACTORING INDUSTRIES	ORGANIS
ETEC		ENTERO TOXISCHE ESCHERICHIA COLI	CLINCHEM
ETFE		ETHYLENE TETRAFLUOROETHYLENE COPOLYMER	COMP
ETH		EIDGENOESSISCHE TECHNISCHE HOCHSCHULE (CH)	INSTI
ETH		ETHIONAMIDE	COMP
ETHZ,ETHL		EIDGENOESSISCHE HOCHSCHULE ZUERICH , LAUSANNE (CH)	INSTI
ETIC		ENVIRONMENT TERATOLOGY INFORMATION CENTER (US)	INSTI LIT
ETL		EMITTER-FOLLOWER TRANSISTOR LOGIC	ELECT
ETO	EtO	ETHYLENE OXIDE	COMP
ETS		EMERGENCY TEMPORARY STANDARD (OSHA,US)	NORM
ETSA		(TRIMETHYLSILYL) ACETICACID ETHYLESTER & ACETATE	COMP
ETU		ETHYLENE THIOUREA	COMP
ETV		ELECTROTHERMAL VAPORIZATION & VAPORIZER	METH INSTR
ETYA		EICOSA-5,8,11,14-TETRAIN ACID	COMP
EU		POLYETHER-URETHANE RUBBER	MATER POLYM
EUCHEM		EUROPEAN CHEMISTRY CONFERENCES (EUR)	CONF
EUCMOS		EUROPEAN CONGRESS MOLECULAR SPECTROSCOPY	CONF
EURIPA		EUROPEAN INFORMATION PROVIDERS ASSOCIATION (EUR)	ORGANIS

Abkürzung Akromym	Alternative	Bedeutung	Zuordnung
abbreviation akronym	alternative	meaning	related section
EURISM		EUROPÄISCHE INFORMATIONSSYSTEM FÜR INDUSTIELLE MEDIZIN (EUR)	LIT
EVA		ELECTROTHERMAL VAPORIZATION ANALYSIS	METH
EVA		ETHYLENE VINYL ACETATE	COMP
EVAC		ETHYLENE VINYLACETATE COPOLYMERISATE	MATER
EVK		ETHYL VINYL KETONE	COMP
EW	E.W.	EARTH WATCH PROGRAM	ENVIR PROG
EWF		EUROPEAN WAX FEDERATION (EUR)	ORGANIS
EWPCA		EUROPEAN WATER POLLUTION CONTROL ASSOCIATION (EUR)	ORGANIS
EXAFS		EXTENDED X-RAY ABSORPTION FINE STRUCTURE SPECTROSCOPY	METH
EXRL	Ex-RL	EXPLOSIONSSCHUTZ RICHTLINIE (D)	NORM

Abkürzung Akronym abbreviation akronym	Alternative alternative	Bedeutung meaning	Zuordnung related section
F&E	R&D FuE	FORSCHUNGS- UND ENTWICKLUNGSVORHABEN	PROG
F-2,6-P		FRUCTOSE-2,6-BIPHOSPHAT	COMP
F.BRAS	F.Bras.	FARMACOPEA BRASILIANA	NORM
F.U.		FARMACOPEA UFFIZINALE DELLA REPUPLICA ITALIANA	NORM
FA		GAMMA-FETALES ANTIGEN	CLINCHEM
FA		FURFURYL ALCOHOL	COMP
FAA	GFAA	FURNACE ATOMIC ABSORPTION	INSTR
FAAAS		FELLOWS OF THE AMERICAN ASSOCIATION FOR THE ADVANCEMENT OF SCIENCE (US)	ORGANIS
FAAS		FLAME ATOMIC ABSORPTION SPECTROMETY	METH
FAB		FAST ATOM BOMBARDMENT	INSTR
FABMS		FAST ATOM BOMBARDMEMENT MASS SPECTROMETRY	METH
FACC		FOOD ADDITIONS AND CONTAMINANTS COMMITTEE (US)	ORGANIS
FACS		FEDERATION OF ASIATIC CHEMICAL SOCIETIES (INT)	ORGANIS
FACSS		FEDERATION OF ANALYTICAL CHEMISTRY AND SPECTROCOPY SOCIETIES	ORGANIS
FACTS		FRANKLIN ADVISORY CENTRE FOR TOXIC SUBSTANCES (US)	INSTI
FAD		FLAVINE-ADENINE DINUCLEOTIDE	COMP
FAD		FACHAUSSCHUß "DRUCKBEHÄLTER" DES HAUPTVERB. BERUFSGENOSSENSCHAFTEN(D)	ORGANIS
FAES		FLAME ATOMIC EMISSION SPECTROSCOPY	METH
FAFS		FLAME ATOMIC FLUORESCENCE SPECTROSCOPY	METH
FAI		FREE ANDROGEN INDEX	CLINCHEM
FAL		FORSCHUNGSANSTALT FUER LANDWIRTSCHAFT (D)	INSTI
FAM		N-3-FLUORANTHYL MALEIMIDE	COMP
FAME		FATTY ACID METHYL ESTER	COMP
FAMH		FOEDERATIO ANALYTICORUM MEDICINALICUM HELVETICORUM (CH)	ORGANIS
FAMH		FOEDERATIO ANALYTICUM MEDICINALIUM HELVETIORUM (CH)	ORGANIS
FAMOS		FLOATING AVALANCHE INJECTION METAL OXIDE SEMICONDUCTOR	ELCT

Abkürzung Akromym / abbreviation akronym	Alternative / alternative	Bedeutung / meaning	Zuordnung / related section
FAMSO		METHYL-METHYLSULFINYLMETHYL SULFIDE	COMP
FANES		FLAMELESS NON-THERMAL EXCITATION SPECTROMETRY	METH
FAO		FOOD AND AGRICULTURE ORGANISATION OF THE UNITED NATIONS (INT)	INSTI
FAP		FATTY ACID PATTERN	CLINCHEM
FAR-IR		FAR INFRARED	METH
FAS		FLAME ABSORPTION SPECTROSCOPY	METH
FASEB		FEDERATION OF AMERICAN SOCIETIES FOR EXPERIMENTAL BIOLOGY (US)	ORGANIS
FAST		FORECASTING AND ASSESSMENT OF SCIENCE AND TECHNOLOGY (EUROPE)	ORGANIS
FAT		FACULTY OF AGRICULTURE UNIVERITY TOKYO (JAP)	INSTI
FATIPEC		FEDER.D'ASSOC.DE TECHNICIENS DES INDUSTR. DES PAINTURES,EMAUX ET ENCRE DE L'EUROPE CONTINENTALE	ORGANIS
FBC		CUMULATED FOLATE BINDING CAPACITY	CLINCHEM
FBE		FLAT BED ELECTROPHORESIS	METH
FBEM		FIXED BEAM ELECTRON MICROSCOPE	INSTR
FBP		FOLATE BINDING PROTEIN	CLINCHEM
FBP		FRUCTOSE BIS PHOSPHATE	COMP
FC		FLASH CHROMATOGRAPHY	METH
FCC		FLUID CARCKING CATALYSTS & FLUID CATALYTIC CRACKING	TECHN
FCC		FOOD CHEMICAL CODEX	NORM
FCCP		TRIFLUORO METHOXY CARBONYL CYANIDE HYDRAZONE (ENZYMINHIB)	COMP CLINCHEM
FCET		FLOW CYTOMETRIC ENERGY TRANSFER MEASUREMENT	CLINCHEM METH
FCIO		FACHVERBAND DER CHEMISCHEN INDUSTRIE OESTERREICHS	ORGANIS
FCKW		FLUOR-CHLOR KOHLENWASSERSTOFFE	COMP
FCM		FLOW CYTOMETRY	CLINCHEM
FCN		FEDERATIE DER CHEMISCHEN NYVERHEID (BELG)	ORGANIS
FCST		FEDERAL COUNCIL FOR SCIENCE AND TECHNOLOGY (US)	INSTI
FD		FIELD DESORPTION	INSTR
FD		FLASH DESORPTION	INSTR

Abkürzung Akromym	Alternative	Bedeutung	Zuordnung
abbreviation akronym	alternative	meaning	related section
FD&C		FOOD, DRUG AND COSMETIC ACT (US)	REG
FDA		FOOD AND DRUG ADMINISTRATION (US)	INSTI
FDCD		FLUORESCENCE DETECTED CIRCULAR DICHROISM	INSTR
FDG	2-FDG	DEOXY-2-FLUORO-D-GLUCOSE	COMP CLINCHEM
FDKI		FORENIGEN DANSKE KEMISKE INDUSTRIER (DK)	ORGANIS
FDM		FIELD DESORPTION MICROSCOPY	METH
FDMA		PERFLUORO-N,N-DIMETHYLCYCLOHEXYLMETHYLAMINE	COMP
FDMS		FIELD DESORPTION MASS SPECTROMETRY	METH
FDN		FLAVIN DI NUCLEOTIDE	COMP
FDNB	DNF	1-FLUORO-2,4-DINITRO BENZENE	COMP REAG
FDNDEA		5-FLUORO-2,4-DINITRO-N,N -DIETHYLANILINE	COMP
FDP		D-FRUCTOSE-1,6-DIPHOSPHATE	COMP
FDS		FIELD DESORPTION SPECTROSCOPY	METH
FDU		FREEZE DRYING UNIT	INSTR
FE	F&E, R&D	FORSCHUNG UND ENTWICKLUNG	PROG
FEA		FEDERATION EUROPEENNE DES ASSOCIATIONS D'AEROSOLS (EUR)	ORGANIS
FEA		FEDERAL ENERGY ADMINISTRATION (US)	INSTI
FEB		FEDERATIONS DES ENTERPRISES DE BELGE (BELG)	ORGANIS
FEBI		FEDERATIONS EUROPEENNES PARBRANCHE D'INDUSTRIES (UNICE,BRÜSSEL)	ORGANIS
FEBS		FEDERATION OF EUROPEAN BIOCHEMICAL SOCIETIES (EUR)	ORGANIS
FECC		FEDERATION EUROPEENNE DU COMMERCE CHIMIQUE (EUR)	ORGANIS
FECS		FEDERATION OF EUROPEAN CHEMICAL SOCIETIES (EUR)	ORGANIS
FED		FIELD EFFECT DIODE	ELCT
FEE		FIELD ELECTRON ENERGY SPECTROSCOPY	METH
FEEC		FEDERATION EUROPEENNE DU COMMERCE CHIMIQUE (EUROPE)	ORGANIS
FEEC		FACULTATIV ENTEROPHATHOGENE ESCHERICHIA COLI	CLINCHEM
FEED		FIELD EMISSION ENERGY DISTRIBUTION	METH
FEEM		FIELD ELECTRON EMISSION MICROSCOPY	METH

Abkürzung Akromym	Alternative	Bedeutung	Zuordnung
abbreviation akronym	alternative	meaning	related section
FEEM		FEDERATION OF EUROPEAN EXPLOSIVES MANUFACTURERS (CEFIC)	ORGANIS
FEES		FIELD ELECTRON ENERGY SPECTROSCOPY	METH
FEFANA		FEDERATION EUROPEENNE DES FABRICANTS D'ADJUBANTS POUR LA NUTRITION ANIMALE (BONN)	ORGANIS
FEIA		FLUORESCENCE-ENZYME-IMMUNO ASSAY	METH
FEICA		FEDERATION EUROPEENNE DES INDUSTRIES DE COLLES ET ADHESIFS (DÜSSELDORF)	ORGANIS
FEIQUE		FEDECION EMPRESARAIL DE LA INDUSTRIA QUIMICA ESPANOLA (SP)	ORGANIS
FEL		FREE ELECTRON LASER	INSTR
FELASA		FEDERATION OF EUROPEAN LABORATORY ANIMAL SCIENCE ASSOCIATION (EUR)	ORGANIS
FEM		FIELD EMISSION MICROSCOPY	METH
FEM		FINITE ELEMENT METHOD	MATER
FEP		FLUORINATED ETHYLENE PROPYLENE COPLYMER (=TETRAFLUOROETHYLEN- PERFLUOROPROPYLEN)	MATER POLYM
FES		FIELD EMISSION SPECTROSCOPY	METH
FES		FLAME EMISSION SPECTROSCOPY	METH
FESM		FEDERATION OF EUROPEAN SOCIETIES OF MICROBIOLOGY	ORGANIS
FET		FIELD EFFECT TRANSITOR	ELECT INSTR
FET		HIGHWAY FUEL ECONOMY TEST (US)	REG
FETI		FLUORESCENCE ENERGY TRANSFER IMMUNO ASSAY	METH
FFA		FREE FATTY ACID	COMP
FFAP		FREE FATTY ACID PHASE	MATER
FFC		FREE FROM CHLORINE	MATER
FFF		FIELD FLOW FRACTIONATION (SEDIMENTATION)	METH
FFH		FORM RAHMEN FILTER HAUBEN (TIERH)	BIOS
FFS		FLAME FLUORESCENCE SPECTROSCOPY	METH
FFT		FAST FOURIER TRANSFORM	METH
FFTS		FAST FOURIER TRANSFORM SPECTROMETRY	METH
FGGE		FIRST GLOBAL RESEACHEPROGRAMME GLOBAL EXPERIMENT	PROG
FGK		FORSCHUNGSGESELLSCHAFT KUNSTSTOFFE (DKI ,D)	ORGANIS

Abkürzung Akromym abbreviation akronym	Alternative alternative	Bedeutung meaning	Zuordnung related section
FH		FACHHOCHSCHULE (D)	INSTI
FH-THEOR		FLORY-HUGGINS-THEORY	POLYM THEOR
FHC	FHC's	FLUORO HYDROCARBONS	COMP
FHG	FhG	FRAUENHOFER GESELLSCHAFT (D)	INSTI
FHTS		FAST HADAMAND TRANSFER SPECTROSCOPY	METH
FHZ		FERRITIN HYDRAZIDE	COMP
FI		FIELD IONIZATION	INSTR
FI	F.I.	FAERBUNGSINDEX	CLINCHEM
FI		FORMALDEHYDE INSTITUTE (SCARSDALE,US)	COMPANY
FIA		FLOW INJECTION ANALYSIS	METH
FIA		FLUORECENCE IMMUNO ASSAY	METH
FIA		FLUORESCENCE INDICATOR ADSORPTION	METH
FIAT		FIELD IMFORMATION AGENCY - TECHNICAL (US)	INSTI
FIB		FAST ION BOMBARDMENT (MASS SPECTROMETRY)	INSTR
FIB		FORSCHUNGSINSTITUT BORSTEL (D)	INSTI
FIC		FIELD IONIZATION CINETICS	THEOR
FIC		FAST ION CHROMATOGRAPHY	METH
FICB		FEDERATION INDUSTRIE DES CHEMIQUE DE BELGIQUE (BELG)	ORGANIS
FICI		FEDERATION OF IRISH CHEMICAL INDUSTRIES (DUBLIN)	ORGANIS
FID		FLAME IONIZATION DETECTOR	INSTR
FID		FREE INDUCTION DECAY	INSTR
FIDOH		FLAME IONIZATION DETECTOR OXYGEN-HYDROGEN	INSTR
FIFE		FEDERATION INTERN.DES ASSOC. DES FABRICANTSDES PRODUIT D'ENTRETIEN	ORGANIS
FIFO		FIRST IN FIRST OUT	ELECT
FIIM	IFPMA	FEDEDATION INTERNATIONALE DE L'INDUSTRIES DES MEDICAMENTS	ORGANIS
FIM		FIELD ION MICROSCOPY	METH
FIMS		FIELD IONIZATION MASS SPECTROMETRY	METH
FIP		FEDERATION INTERNATIONALE PHARMACEUTIQUE (INT)	ORGANIS
FIR	FAR-IR	FAR INFRARED	METH
FIS		FIELD IONIZATION SPECTROSCOPY	METH
FIT		FLOW INJECTION TITRIMETRY	METH

Abkürzung Akromym abbreviation akronym	Alternative alternative	Bedeutung meaning	Zuordnung related section
FITC		FLUORESCEIN ISOTHIO CYANATE	COMP
FIW		FORSCHUNGSINSTITUT FÜR WASSERTECHNOLOGIE (AACHEN)	INSTI
FIZ		FACHINFORMATIONS ZENTRUM (D)	INSTI LIT
FKMS	FMS	FESTKOERPER MASSENSPEKTOMETRIE	METH
FKW	FHC	FLUOR KOHLENWASSERSTOFFE	COMP
FL		FLUORESCENCE	INSTR
FLA		FLUORESCAMIN	COMP
FLA		FILTER LENS ANALYZER	INSTR
FLAQ		FEDERATION LATIN AMERCAN ASSOCIATI QUICA	ORGANIS
FLC		FERROELECTRIC SMECTIC LIQUID CRYSTAL(S)	MATER
FLU		FLUANTHRENE	COMP
FM		FREQUENCY MODULATION = FREQUENCY MODULATED	INSTR
FMA		FLUORESCEIN MERCURIC ACETATE	COMP
FMH		FOEDERATIO MEDICORUM HELVETICORUM (CH)	ORGANIS
FMI		FRIEDRICH MISCHER INSTITUT (BASEL)	INSTI
FMN		FLAVIN MONO NUCLEOTIDE	CLINCHEM
FMOC-CL	FMCO-Cl	9-FLURENYLMETHOXY CARBONYL CHLORID (=9-FLUORRENYLMETHYL-CHLOROFORMATE)	COMP REAG
FMQ		FLUORINATED METHYLSILICONE RUBBER	MATER
FMS	FKMS	FESTKOERPER MASSEN SPECTOMETRIE	METH
FNBS		4-FLUORO-3-NITROBENZENE SULFONIC ACI	COMP REAG
FNM	FN M	FACHNORMENAUSSCHUß MATERIALPRÜFUNG (D)	NORM
FNP	FNPyr	2-FLUORO-3-NITRO PYRIDINE	COMP REAG
FNPS		BIS (4-FLUORO-3-NITROPHENYL) SULFONE	COMP
FNPS	FN PS	FACHNORMENAUSSCHUß PERSÖNL SCHUTZAUSRÜST.& SICHERHEITSKENNZEICHNUNG (D)	NORM
FNU	TUF	FORMAZINE NEPHELOMETRIC UNITS	EVAL
FNW	FN W	FACHNORMENAUSSCHUß WASSERWESEN (D)	NORM
FOA		FORSCHUNGSAUSSCHUß (EC)	INSTI
FOC		FACHVEREINIGUNG ORGANISCHE CHEMIE (VCI,D)	ORGANIS
FOD		PERDEUTERO HEPTAFLUORO-OCTAN-DIONATO-3,5	COMP

Abkürzung Akromym	Alternative	Bedeutung	Zuordnung
abbreviation akronym	alternative	meaning	related section
FOLZ		FIRST ORDER LAUE ZONE	SPECT
FORTRAN		FORMULAR TRANSLATION	ELECT
FOTL		FIBRE OPTIC THIN LAYER	INSTR
FP		GAMMA-FETO-PROTEIN	CLINCHEM
FPD		FLAME PHOTOMETRIC DETECTOR	INSTR
FPIA		FLUORESCENCE POLARIZATION IMMUNO ASSAY	METH
FPLA		FIELD PROGRAMABLE LOGIC ARRAY	ELECT
FPLC		FAST PROTEIN, PEPTIDE AND POLYNUCLEOTIDE LIQUID CHROMATOGRAPHY	METH
FPM		VINYLIDENFLUORIDE HEXAFLUOROPROPYLEN COPOLYMER	MATER
FPR		FLUORESCENCE PHOTOBLEACHING RECOVERY	METH
FPRL		FOREST PRODUCTS RESEARCH LABORATORY (MADISON,US)	INSTI
FPROM		FIELD PROGRAMABLE READ ONLY MEMORY	ELECT
FR		FORBIDDEN REFLECTION	SPECT
FRA		FETO RENALES ANTIGEN	CLINCHEM
FRAME		FUND FOR REPLACEMENT OF ANIMALS IN MEDICAL EXPERIMENTS (GB)	INSTI
FRCA		FLAME RETARDANT CHEMICAL ASSOCIATION	ORGANIS
FRG	BRD , D	FEDERAL REPUBLIC OF GERMANY (D)	
FRH		FOLLICEL STIMULATING RELEASING HORMONE	CLINCHEM
FRI		FOOD RESEARCH INSTITUTE (TOKYO,JAP)	INSTI
FROM		FACTORY PROGRAMABLE READ ONLY MEMORY	ELECT
FS		FUSED SILICA	MATER
FS		FERMI SURFACE	SPECT
FS		FREMY'S SALT = DIPOTASSIUM NITROSODISULFONATE	COMP
FSA		FETALES SULPHOGLYCOPROTEIN	CLINCHEM
FSC		FOOD STANDARDS COMMITTEE (US)	ORGANIS NORM
FSC		FUSED SILICA CAPILLARY	INSTR
FSCGC	FSC-GC	FUSED SILICA CAPILLARY GAS CHROMATOGRAPHY	METH
FSF		FIBRINSTABILZING FACTOR	CLINCHEM
FSGO		FLOATING SPHERICAL GAUSSIAN ORBITAL	THEOR

Abkürzung Akromym abbreviation akronym	Alternative alternative	Bedeutung meaning	Zuordnung related section
FSH		FOLLICLE-STIMULATING HORMONE	CLINCHEM
FSH-RH		FOLLISERIN RELEASING HORMONE	CLINCHEM
FSK		FACHVERBAND SCHAUMKUNSTSTOFFE (D)	ORGANIS
FSOT		FUSED SILICA OPEN TUBULAR CAPILLARY	INSTR
FSR		FREE SPECTRAL RANGE	SPECT
FSS		FACHVERBAND FÜR STRAHLENSCHUTZ (D)	ORGANIS
FT		FOURIER TRANSFORM	INSTR
FT		FISCHER TROPSCH	TECHN
FT-ESR		FOURIER TRANSFORM ELECTRON SPIN RESONANCE	METH
FTA		FLOCCULATION TEST APPARATUS	INSTR
FTA-ABS		FLUORESCENT TREPONEMAL ANTIBODY ABSORBTION	CLINCHEM
FTC		FEDERAL TRADE COMMISSION (US)	INSTI
FTI		FREE THYROXIN INDEX	CLINCHEM
FTICR		FOURIER TRANSFORM ION CYCLOTRON RESONANCE	METH
FTIR		FOURIER TRANSFORM INFRARED SPECTROSCOPY	METH
FTMS	FT-MS	FOURIER TRANSFORM MASS SPECTROMEY	METH
FTN		PERFLUORO-1,3,7-TRIMETHYL BICYCLO(3.3.1)-NONANE	COMP
FTNQR	FT-NQR	FOURIER TRANSFORM NUCLEARE QUADRUPOL RESONANCE	METH
FTS		FOURIER TRANSFORM SPECTROSCOPY/METRY	METH
FTU		FORMAZIN TURBIDITY UNIT	EVAL
FTZ		FERNMELDE TECHNISCHE ZULASSUNG (D)	INSTR REG
FU	F.U.	FARMACOPEA UFFICIALE DELLA REPUBLICA ITALIANA (I)	NORM
FUDR		5-FLUORO-2-DEOXY-BETA-URIDINE	COMP
FUETAP		FORMED UNDER ELEVATED TEMPERATURES AND PRESSURES	TECHN
FVC		FACHGRUPPE VERFAHRENS UND CHEMIEINGENIEUR TECHNIK SIA (CH)	ORGANIS
FW		FACHGRUPPE WASSERCHEMIE (GDCH,D)	ORGANIS
FWHH	FWHM	FULL WIDTH OF HIGHT OF A SPECTRAL PEAK	SPECT
FWHM	FWHH	FULL WIDTH OF HALF MAXIMUM OF A SPECTRAL PEAK	SPECT
FWTM		FULL WIDTH OF A TENTH MAXIMUM	INSTR SPECT

Abkürzung Akromym	Alternative	Bedeutung	Zuordnung
abbreviation akronym	alternative	meaning	related section
FXD		FLASH X-RAY DIFFRACTION	METH
FXR.		FLASH X-RAY RADIOGRAPHY	METH
FYU		DEPARTMENT OF FERMENTATION YAMANISHI UNIVERSITY (JAP)	INSTI

Abkürzung Akromym	Alternative	Bedeutung	Zuordnung
abbreviation akronym	alternative	meaning	related section
G-6-P		GLUCOSE-6-PHOSPHATE	COMP
G-6-PDH		GLUCOSE-6-PHOSPHATE DEHYDROGENASE	CLINCHEM
GA3		GIBBERELIC ACID	CLINCHEM
GAA		GLACIAL ACETIC ACID	COMP
GAA		GUANIDINO ACETIC ACID	COMP
GAA		GEWERBEAUFSICHTSAMT (D)	INSTI
GABA		GAMMA-AMINOBUTYRIC ACID =4-AMINOBUTYRIC ACID	COMP
GABOB		GAMMA AMINO-BETA-HYDROXY BUTYRIC ACID	COMP
GAC		GRANULAR ACTIVATED CARBON	MATER
GAC		GAS ADSORPTION CHROMATOGRAPHY	METH
GAMS		GROUPEMENT POUR L'AVANCEMENT DES METHODES SPECTROSCOPIQUES ET PHYSICOCHIMIQUES D'ANALYSE	ORGANIS
GAP		GOOD AGRICULTURE PRACTICE	BIOS
GAPGDH		GLYCERALDEHYDE-3'PHOSPHATE DEHYDROGENASE	CLINCHEM
GARP		GLOBAL ATMOSPHERIC RESEARCH PROGRAM	PROG
GASFET		GASSENSITIVE FIELD EFFECT TRANSISTOR	INSTR
GATE		GARP ATLANTIC TROPICAL EXPERIMENT	ENVIR
GATT		GENERAL AGREEMENT ON TRAFICS AND TRADE (UN)	REG
GAU		GROESSTER AUSZURECHNENDER UNFALL	TECHN
GBCH		GESELLSCHAFT F BIOLOGISCHE CHEMIE (D)	ORGANIS
GBF		GESELLCHAFT F BIOTECHNOLOGISCHE FORSCHUNG (D)	INSTI
GBHA		GLYOXAL-BIS(2-HYDROXY ANILE)	COMP
GC		GLASSY CARBON	MATER
GC		GASCHROMATOGRAPHY	METH
GC2		GLAS CAPILLARY GASCHROMATOGRAPHY	METH
GCE		GLASSY CARBON ELECTRODE	INSTR E-CHEM
GDC		VDI-GESELLSCHAFT VERFAHRENSTECHNIK UND CHEMIEINGENIEURWESEN (D)	ORGANIS
GDCH	GDCh	GESELLSCHAFT DEUTSCHER CHEMIKER (D)	ORGANIS
GDCP		GESELLSCHAFT FUER DIDAKTIK DER CHEMIE UND PHYSIK (D)	ORGANIS
GDD		GASDISCHARGE DISPLAY	INSTR
GDL		GLOW DISCHARGE LAMP	INSTR

Abkürzung Akromym	Alternative	Bedeutung	Zuordnung
abbreviation akronym	alternative	meaning	related section
GDMB		GESELLSCHAFT DEUTSCHER METALLHUETTEN- UND BERGPROBENEHMER (D)	ORGANIS
GDMS		GLOW DISCHARGE MASS SPECTROMETRY	METH
GDNAE		GESELLSCHAFT DEUTSCHER NATURWISSENSCHAFTLER UND AERZTE (D)	ORGANIS
GDOS		GLOW DISCHARGE OPTICAL SPECTROSCOPY	INSTR
GDP		GUANOSINE-5'-DIPHOSPHATE	COMP
GDR	DDR	GERMAN DEMOCRATIC REPUBLIC	
GDS		GRAPHIC DATA SYSTEM	ELECT
GE		GASENTHALPIMETRIE	METH
GEFAB		GROUPEMENT EUROPEEN DES ASS NATION. DE FARBICANTS DES PESTICIDES	ORGANIS
GEFAP		GROUPMENT EUROPEEN DES ASSOC. NATIONALES DES FABRICANTS DE PESTICIDES (BRÜSSEL)	ORGANIS
GEFTA		GESELLSCHAFT FÜR THERMISCHE ANALYSE (D)	ORGANIS
GELI		GEMANIUM-LITHIUM SEMICONDUCTOR DETECTOR	INSTR
GEMCQ		GIDDINGS,EYRING EXTENDED BY MC QUARRIN	THEOR
GEMS		GLOBAL ENVIRONMENTAL MONITORING SYSTEM (WHO,UNEP)	ENVIR
GEP		GENERAL ELECTRIC PLASTICS	COMPANY
GEP		GASTRO ENTERO PANKREATISCHES SYSTEM	CLINCHEM
GERM		GROUPE D'ETUDES EN RESONANCE MAGNETIQUE (F)	ORGANIS
GESA		GESELLSCHAFT EXPERIMENTELLE SPANNUNGSANALYSE (VDI,VDE,D)	ORGANIS
GESAMP		JOINT GROUP OF EXPERTS ON SCIENTIFIC ASPECTS OF MARINE POLLUTION	ORGANIS
GESSG		GESUNDHEITS-SICHERSTELLUNGS GESETZ (D)	REG
GFA		GLIAL FIBRILLARY ACIDIC PROTEIN	CLINCHEM
GFAAS		GRAPHITE FURNACE ATOMIC ABSORPTION SPECTROSCOPY	INSTR
GFAH	GfAH	GESELLSCHAFT FÜRARBEITSSCHUTZ-UND HUMANISIERUNGSFORSCHUNG (DORTMUND)	ORGANIS

Abkürzung Akromym	Alternative	Bedeutung	Zuordnung
abbreviation akronym	alternative	meaning	related section
GFC		GEL FILTRATION CHROMATOGRAPHY	METH
GFC	GFK	GLAS FIBRE CARBON	MATER
GFK	GFC	GLASFASERVERSTAERKTER KOHLENSTOFF	MATER
GFK	GfK	GESELLSCHAFT FÜR KERNFORSCHUNG (D)	ORGANIS
GFMD		GOLD FILM MERCURY DETECTOR	ECHEM INSTR
GFR		GLOMERULAR FILTRATION RATE	CLINCHEM
GFS		GESELLSCHAFT F SICHERHEITSWISSENSCHAFT (D)	ORGANIS
GFS		GEMEINSAME FORSCHUNGSSTELLE (EG)	INSTI
GGBS		GROUPEMENT INTERN.DES ASSOC.NATIONALES DES FABRICANTS DE PESTICIDES	ORGANIS
GGE		GROUP OF GOVERNMENT EXPERTS (US)	ENVIRON
GGG		GADOLINIUM GALLIUM GRANAT	MATER
GGTP		GAMMA-GLUTAMYL TRANS PEPTIDASE	COMP
GGVE		GEFAHRENGUT VERORDNUNG EISENBAHN (D)	REG
GGVS		GEFAHRENGUT VERORDNUNG STRASSE (D)	REG
GH		GROWTH HORMON	CLINCHEM
GHP	GMP	GUTE HERSTELLUNGS PRAXIS	TECHN
GHRF	GH-RF GRF	GROWTH HORMONE RELEASING FACTOR	CLINCHEM
GHRIF		GROWTH HORMONE RELEASING INHIBITING FACTOR	CLINCHEM
GIFAP		GROUPEM.INTERN.DES ASSOC.NATION.DE FABRICANTS DE PROD. AGROCHIMIQUES	ORGANIS
GIH		GROWTH HORMONE INHIBITING HORMONE	CLINCHEN
GIIP		GROUPEMENT INTERN. DE L'INDUSTRIE PHARMACEUTIQUE DES PAYS EUROPEENS (BRÜSSEL)	ORGANIS
GILSP		GOOD INDUSTRIAL LARGE SCALE PRACTICE	TECHN
GIM		GENETICS OF INDUSTRIAL MICROORGANISM	CLINCHEM
GIP		GASTRIC INHIBITING POLYPEPTIDE	COMP
GIPME		GLOBAL INVESTIGATION OF POLLUTION IN THE MARINE ENVIRONMENT	ENVIR
GIR		GLANCING INCIDENCE REFLECTION	INSTR
GIT		GASTRO INTESTINAL TRACT	CLINCHEM
GITC		GLUCOPYRANOSYL ISOTHIOCYANATE	COMP
GIV		GRAPHIT INTERACTION VERBINDUNG	COMP
GKC		GESELLSCHAFT DEUTSCHER KOSMETIK-CHEMIKER E.V.	ORGANIS

Abkürzung Akromym	Alternative	Bedeutung	Zuordnung
abbreviation akronym	alternative	meaning	related section
GKE	SCE	GESAETTIGTE KALOMEL ELEKTRODE	ECHEM INSTR
GKSS		GESELLS. F. KERNENERGIEVERWERTUNG IN SCHIFFBAU UND SCHIFFAHRT (D)	INSTI
GKV		GESAMTVERBAND KUNSTSTOFFVERARBEITENDE INDUSTRIE (D)	ORGANIS
GKW		GESAMTKOERPERGEWICHT	CLINCHEM
GLC		GAS LIQUID CHROMATOGRAPHY	METH
GLDH		GLUTAMATDEHYDROGENASE	CLINCHEM
GLGA		GINZBURG, LONDON, GARKOW, ABRIKOSOW (SUPRACONDUCTION)	THEOR
GLN	Gln	GLUTAMINE	COMP
GLP		GOOD LABORATORY PRACTICE	METH
GLP		GLUCAGON LIKE PEPTIDE	CLINCHEM
GLU	Glu	GLUTAMIC ACID	COMP
GLUPHEPA		N-GLUTARYL-L-PHENYLALANIN-4-NITROANALID (GLUHAMATE-DEHYDROGENASE)	COMP
GLY	Gly	GLYCINE	COMP
GLYMO		3-GLYCIDYLOXYPROPYLTRIMETHOXYSILANE	COMP
GM		GEIGER-MÜLLER ZÄHLROHR	INSTR
GM-CSF		GRANULOCYTE/MACROPHAGE COLONY STIMULATING FACTOR	CLINCHEM
GMD		GESELLSCHAFT F MATHEMATHIK UND DATENVERARBEITUNG (D)	INSTI
GMDS		DEUTSCHE GESELLSCHAFT MEDIZINISCHE DOKUMENTATION,INFORMATION & STATISKTIK	ORGANIS
GMP	cGMP	GUANOSINE 3',5'-CYCLIC MONOPHOSPHATE	COMP
GMP	GHP	GOOD MANUFACTURING PRACTICE	TECHN
GMR		VDI/VDE - GESSELLSCHAT MEß-UND REGELUNGSTECHNIK (D)	ORGANIS
GMS		GLYCERINMONOSTERAT	COMP
GMT		GLASMATTENVERSTAERKTE THERMOPLASTE	MATER
GMT		GOOD MICROBIOLOGICAL TECHNIQUES	CLINCHEM
GMU		GESELLSCHAFT F MATERIALRUECKGEWINNUNG UND UMWELTSCHUTZ (D)	COMPANY
GMV		GUARANTEED MINIMUM VALUE	COMP
GNP		GROSS NATIONAL PRODUCT /BRUTTO SOZIAL PRODUKT	

Abkürzung Akromym	Alternative	Bedeutung	Zuordnung
abbreviation akronym	alternative	meaning	related section
GOD		GENERATION OF DISVERSITY	CLINCHEM
GOD		GLUCOSE OXIDASE	CLINCHEM
GOFO		GEOMETRIC OPTIMATED FLOATING ORBITALS	THEOR
GOI		GAIN ON IGNITION (X-RAY)	SPECT
GOST		USSR STANDARDS	NORM
GOT		GLUTAMT OXALACETAT TRANSAMINASE	COMP
GOT/GPT		DE-RITIS-QUOTIENT	CLINCHEM
GPC		GEL PERMEATION CHROMATOGRAPHY	METH
GPF		GAS PHASE FLUORESCENCE	METHOD
GPIB		GENERAL PURPOSE INTERFACE BUS	ELECT
GPMAS		GAS PHASE MOLECULAR ABSORPTION SPECTROSCOPY	METH
GPO		ALLYLGLYCIDE-ETHER-PROPYLENEOXIDE COPOLYMER	MATER
GPPS		GENERAL PURPOSE SIMULATION SYSTEM	THEOR
GPRMC		GROUPMENT DES PLASTIQUES RENFORCES DU MARCHE COMMUN (CH)	ORGANIS
GPT		GAS-PHASE TITRATION	METH
GPT		GLUTAMAT-PYROVAT-TRANSAMINASE	COMP
GR		GUARANTEED REAGENT	COMP
GRAPS		GROUPE DE RECHERCHE APPLIQUATION A LA PHYSIK DES SURFACES (CAN)	ORGANIS
GRAS		GENERALLY REGARDED AS SAFE	COMP ENVIR
GRF		GROWTH HORMONE RELEASING FACTOR	CLINCHEM
GRH		GROWTH RELEASING HORMONE =STH-RELEASING HORMONE	CLINCHEM
GRID		GLOBAL RESOURCE INFORMATION DATABASE (UNO,INT)	LIT ENVIR
GRS		GESELLSCHAFT F REAKTORSICHERHEIT (D)	ORGANIS
GS		GESCHUETZTE SICHERHEIT (D)	NORM
GSA		GUANIDINO SUCCINIC ACID	COMP
GSAM		GENERALIZED STANDARD ADDITION METHOD	METH
GSASA		GESELLSCHAFT SCHWEIZER. AMTS- UND SPITALAPOTHEKER (CH)	ORGANIS
GSB		GESELLSCHAFT ZUR BESEITIGUNG VON SONDERMUELL IN BAYERN (D)	COMPANY
GSC		GAS SOLID CHROMATOGRAPHY	METH
GSCHG	GSchG	GEWÄSSER SCHUTZ GESETZ (CH)	REG

Abkürzung Akromym	Alternative	Bedeutung	Zuordnung
abbreviation akronym	alternative	meaning	related section
GSCL		GAS SOLID CHEMOLUMINESCENCE	METH
GSD		GEL SLAB DRIER	INSTR
GSF		GESELLSCHAFT F STRAHLEN UND UMWELTFORSCHUNG (MÜNCHEN,D)	INSTI
GSG		GESELLSCHAFT SCHWEIZERISCHER GIFTINSPEKTOREN (CH)	ORGANIS
GSGG		GALLIUM, SCANDIUM, GADOLINIUM GRANAT	MATER
GSH		GLUTHATHIONE - STIMULATING HORMONE	CLINCHEM
GSH		GLUTHATHIONE , REDUCED	COMP
GSI		GESELLSCHAFT F SCHWERIONENFORSCHUNG (DARMSTADT,D)	INSTI
GSI		GRAND SCALE INTEGRATION	ELECT
GSIA		GESELLSCHAFT SCHWEIZERISCHER INDUSTRIE APOTHEKER (CH)	ORGANIS
GSSG		GLUTHATHIONE ,OXIDIZED	COMP
GT		GAMMA-GLUTARYL-TRANSPEPTIDASE	CLINCHEM
GTA		GRAPHITE TUBE ATOMIZER	INSTR
GTCP		GEWERKSCHAFT TEXTIL CHEMIE KERAMIK (D)	ORGANIS
GTD		GAUSS TYPE DISTRIBUTION	THEOR
GTN		GLYCEROL TRINITRATE	COMP
GTO		GAUSS TYPE ORBITAL	THEOR
GTP		GUANOSINE-5'-TRIPHOSPHATE	COMP
GTS		GASTIGHT SYRINGE	INSTR
GTU		GESELLSCAFT FÜR TECHNISCHE ÜBERWACHUNG	ORGANIS
GTZ		DEUTSCHE GESELLSCHAFT F TECHNISCHE ZUSAMMENARBEIT (D)	INSTI
GUGA		GRAPHICAL UNITARY GROUP APPROACH	THEOR
GUM		GESELLSCHAFT FÜR UMWELT-MUTATIONSFORSCHUNG E.V.	ORGANIS
GUV		GIANT UNILAMILLAR VESICLES	CLINCHEM
GVC		VDI GESELLSCHAFT VERFAHRENSTECHNIK UND CHEMIEINGENIEURWESEN (D)	ORGANIS
GVC		GESELLSCHAFT VERFAHRENSTECHNIK CHEMIEINGENIEURWISSENSCHAFT (CH)	ORGANIS
GVH	GvH	GRAFT-VERSUS-HOST DISEASE	CLINCHEM
GVT		FORSCHUNGS-GESELLSCHAFT VERFAHRENS TECHNIK (D)	ORGANIS
GYEP		GLUCOSE,YEAST EXTRACT PEPTONE	CLINCHEM

Abkürzung Akromym abbreviation akronym	Alternative alternative	Bedeutung meaning	Zuordnung related section
H6CDD	PCDD6	HEXACHLORODIBENZO-P-DIOXIN	COMP
H6CDF	PCDF6	HEXACHLORO-DIBENZO-FURAN	COMP
H7CDD	PCDD7	HEPTACHLORO-DIBENZO-P-DIOXIN	COMP
H7CDF	PCDF7	HEPTACHLORO-DIBENZO-FURAN	COMP
HAB		HOMOEOPATHISCHES ARZNEIBUCH	LIT PHARM
HAB		P-HYDROXY AZOBENZENE	COMP
HABA		2(4'-HYDROXY-PHENYLAZO)BENZOIC ACID	COMP
HABBA		2-(4'HYDROXYAZOBENZENE)BENZOIC ACID HEMOGLOBIN	COMP CLINCHEM
HAD		HEREDITARY ANTITHROMBIN DEFICIENCY	CLINCHEM
HAFP	hAFP	HUMAN ALPHA FETOPROTEIN	CLINCHEM
HAGABA		HYDROXAMACID GAMMA-AMINO-BUTYRIC ACID	COMP
HALS		HINDERED AMINE LIGHT STABILIZERS	MATER
HAN		HEAVY AROMATIC NAPHTHA (OIL)	MATER
HAN		HYBRID ALIGNED NEMATIC CELLS	TECHN
HAP		HYDRATED ANTIMONY PENTOXIDE	COMP
HAPUG		MODULATIONCIRCUIT TO HARLICH,PUNKS UND GERTH	INSTR ELECT
HAT		MIXTURE OF HYPOXANTHIN, AMINOPTERIN AND THYMIDIN	CLINCHEM
HAT		HÄMAGGLUTATIONSTEST	CLINCHEM
HAV		HEPATITS A VIRUS	CLINCHEM
HAW		HIGH ACTIVE WASTE	TECHN
HB	Hb-X	HEMOGLOBIN (X = A,C,F,S)	COMP CLINCHEM
HBA		2-(HEXYLOXY) BENZAMIDE	COMP
HBB		HEXABROMBIPHENYL	COMP
HBCAG	HBcAg	HEPATITS B CORE ANTIGEN	CLINCHEM
HBD		HEXABUTYLDISTANNOXANE	COMP
HBDH		ALPHA-HYDROXY BUTYRATE-DEHYDROGENASE	CLINCHEM
HBEAG	HBeAg	HEPATITS B E ANTIGEN	CLINCHEM
HBF	HbF	FETAL HEMOGLOBIN	CLINCHEM
HBO		2-(2-HYDROXYPHENYL) BENZOXAZOLE	COMP
HBV		HEPATITIS-B-VIRUS	CLINCHEM
HCA		HELVETIA CHEMICA ACTA	LIT
HCCH	HCH,BHC	HEXACHLOROCYCLOHEXANE (PREFER ALTERNATIV)	COMP

Abkürzung Akromym abbreviation akronym	Alternative alternative	Bedeutung meaning	Zuordnung related section
HCDD	PCDD6	HEXACHLORODIBENZO-P-DIOXIN (PREFER ALTERNATIV)	COMP
HCDF	PCDF6	HEXACHLORODIBENZO FURANE (PREFER ALTERNATIV)	COMP
HCG		HUMAN CHORION GONADOTROPIN	CLINCHEM
HCH	BHC,HCCH	HEXACHLOROCYCLOHEXANE	COMP
HCL	HKL	HOLLOW CATHODE LAMP	INSTR
HCS	CGP	HUMAN CHORIONIC SOMATOMAMMO TROPIN	CLINCHEM
HCSA		HALOGENATED CLEANING SOLVENTS ASSOCIATION	ORGANIS
HCT		HUMAN CHORIONIC THYROTROPIN	CLINCHEM
HDBP		BIS(DIBUTYL)PHOSPHORIC ACID	COMP
HDC		HYDRODYNAMIC CHROMATOGRAPHY	METH
HDCBS		2-HYDROCY-3,5-DICHLOROBENZENESULFONIC ACID	COMP
HDDR		HIGH DENSITY DIGITAL MAGNETIC RECORDING (EDV)	ELECT
HDEHP		HYDROGEN-DI(ETHYLCYCLOHEXYL)PHOSPHATE	COMP REAG
HDI	1.6HDI	1.6-HEXANE-DIISOCYANATE	COMP
HDL		HIGH DENSITY LIPOPROTEIN	COMP
HDODA		1,6-HEXANEDIOL DIACRYLATE	COMP
HDPE		HIGH DENSITY POLYETHYLENE	MATER
HDR		HEISSDAMPFREAKTOR	TECHN
HDS		HYDRIER DESULFURING(CATALYST)	COMP
HDV		HEPATITIS- D-VIRUS	CLINCHEM
HEA		HYDROXYLETHYL ACRYLATE	COMP
HEA		N-(2-HYDROXYETHYL) AZIRIDINE	COMP
HECD		HALL ELECTROCONDUCTIVITY DETECTOR	INSTR
HEDS		HYDROXYLETHYLDISULFIDE	COMP
HEDTA		N-HYDOXYETHYLETHYLENEDIAMINE TETRAACETATE	COMP
HEED		HIGH ENERGY ELECTRON DIFFRACTION	METH
HEEI		N-(2-HYDROXYETHYL) ETHYLENEIMINE	COMP
HEHIXE		HIGH ENERGY HEAVY IONS X-RAY EMISSION	METH
HEIS		HIGH ENERGY IONS BACK SCATTERING SPECTROSCOPY	METH
HEL		HIGH ENERGY LASER	INSTR

Abkürzung Akromym abbreviation akronym	Alternative alternative	Bedeutung meaning	Zuordnung related section
HEMA		2-HYDROXYETHYL METHACRYLATE	COMP
HEMPA	HMPT	HEXAMETHYLPHOPHORIC ACID TRIAMIDE (PREFER ALTERNATIVE)	COMP
HEOD		DIELDRINE EXO-ENDO	COMP
HEPA		HIGH EFFICIENCY PARTICULATE AIR FILTER	INSTR
HEPES		2(4-(HYDROXYETHYL)PIPERAZINYL-(1)) ETHANESULFONIC ACID (BUFFER)	COMP
HEPPES		N-2-HYDROXYETHYL PIPERAZINE-N-3-PROPANE SULFONIC ACID (BUFFER)	COMP
HEPSO		N-HYDROXYETHYLPIPERAZINE-N'-2-HYDROXYPROPANESULFONIC ACID	COMP
HERD		HIGH ENERGY X-RAY DIFFRACTION	METH
HES	6-H.E.S	HYDROXYETHYL STARCHES	MATER
HET		HEXAETHYLTETRAPHOSPATE	COMP
HET		HENS EGG TEST	CLINCHEM
HETE	5-HETE	5-HYDROXY-6,8,11,14 EICOSOTETRAENOIC ACID	COMP
HETP		HIGHT EQUIVALENT TO A THEORETICAL PLATE	METH TECHN
HEW		HEALTH,EDUCATION AND WELFARE DEPARTMENT (US)	INSTI
HEXA	HMT	HEXAMETHYLENE TETRAMINE	COMP POLYM
HF		HIGH FREQUENCY	INSTR
HF		HARTREE-FOCK	THEOR
HF-LASER		HYDROGENFLUORIDE-LASER	INSTR
HFA		HEXAFLUOROACETON	COMP
HFB	HFB-	HEPTAFLUOROBUTYRYL LIGAND	COMP
HFBA		HEPTAFLUORO BUTYRIC ACID ANHYDRATED	COMP
HFBI		HEPTAFLUORO BUTYRYL IMIDAZOLE	COMP
HFCS		HIGH FRUCTOSE CORN SYRUP	MATER CLINCHEM
HFI	HFIP	1,1,1,3,3,3-HEXAFLUOROPROPAN-2-OL (=HEXAFLUOROISIPROPANOL)	COMP
HFO		HIGH FREQUENCY OSCILLATOR	INSTR
HFP		HEXAFLUOROPROPENE	COMP
HFS		HYPERE FINE STRUCTURE	METH SPECT
HFSH	hFSH	HUMAN FOLLICLE STIMULATING HORMONE	CLINCHEM
HFTA		HEXAFLUOROTHIOACETONE	COMP
HFTBA		HEPTACOSAFLUOROTRI-N-BUTYLAMINE	COMP

Abkürzung Akromym abbreviation akronym	Alternative alternative	Bedeutung meaning	Zuordnung related section
HGA		HYDROXYL-GROUP ACTIVATION	COMP
HGA	HGAA	HYDRIDE GRAPHITROHRKUEVETTEN ATOM ABSORPTION	METH
HGAA	HGA	HYDRIDE GENERATION FOLLOWED BY ATOMIC ABSORPTION	METH
HGF		HYPERGLYCEMIC-GLYCOGENOLYTIC FACTOR	CLINCHEM
HGH	GH,STH	HUMAN GROWTH HORMONE	CLINCHEM
HGPRT		HYPOXANTHINGUANOSIN-PHOSPHO RIBOSYL TRANSFERASE	CLINCHEM
HHI		HEINRICH HERTZ INSTITUT (BERLIN,D)	INSTI
HHL		HYPOPHYSENHINTERLAPPEN HORMON	CLINCHEM
HHPA		HEXAHYDROPHTHALIC ANHYDRIDE	COMP
HIAA		5-HYDROXYINDOLE-3-ACETIC ACID	COMP
HIC		HYDROPHOBIC INTERACTION CHROMATOGRAPHY	METH
HIC		HYBRID INTEGRATED CIRCUIT	ELECT
HIEF		HYBRID ISOELECTRIC FOCUSSING	INSTR
HIFI		HIGH FIDELITY	ELECT
HIIDMS		HEAVY ION INDUCED MASS SPECTROMETRY	METH
HIM		HYGIENE INSTITUT DER UNIVERSITÄT MARBURG (D)	INSTI
HINIL		HIGH NOISE IMMUNITY LOGIC	ELECT
HIP		HOT ISOSTATIC PRESSING	TECHN
HIPIP		HIGH POTENTIAL IRON PROTEIN	CLINCHEM
HIS	His	HISTIDINE	COMP
HIV	HTLV-III	HUMAN IMMUNODEFICIENCY VIRUS	CLINCHEM
HIXSE		HEAVY ION INDUCED X-RAY SATELLITE EMISSION	METH
HJZ		HYDRIER JODZAHL (ANALYSE FETTE,SEIFEN)	METH
HKL	HCL	HOHLKATHODENLAMPE (PREFER ALTERNATIV)	INSTR
HLA		HEAVY AND LIGHT ACETIC ACID ANTIGEN	CLINCHEM
HLB		HYDROPHIL-LIPOPHIL BALANCE	THEOR
HLFR		HIGH LEVEL FARADAIC RECTIFICATION	INSTR
HLH	hLH	HUMAN LUTEINIZING HORMONE	CLINCHEM
HLT		HELIUM LEAK TESTING	INSTR TECHN
HLW		HIGH LEVEL RADIOACTIVE WASTE	ENVIR
HMA		HEART MUSCLE ANTIBODY	CLINCHEM
HMAC		HAZARDOUS MATERIALS ADVISORY COMMITTEE (US)	INSTI

Abkürzung Akromym	Alternative	Bedeutung	Zuordnung
abbreviation akronym	alternative	meaning	related section
HMAT		HEXA(1-(2-METHYL)AZIRIDINYL)-1,3,5-TRIPHOSPHATRIAZINE	COMP
HMB		2-HYDROXY-4-METHOXYBENZOPHENONE	COMP
HMB		2-HYDROXY-5-METHOXYBENZALDEHYDE	COMP
HMBT		2-HYDRAZO-2.3-DIHYDRO-3METHYLBENZOTHIAZOLE	REAG
HMDAC		HEXAMETHYLENE DIAMINE CARBAMATE	COMP POLYM
HMDE		HANGING MERCURY DROP ELECTRODE	INSTR EANAL
HMDS		1,1,1,3,3,3-HEXAMETHYL DISILAZANE	COMP REAG
HMDSO		HEXAMETHYLDISILOXANE	COMP
HME		9-HYDROXY-2-METHYL ELLIPTICINIUM	COMP
HMF		5-HYDROXY METHYL FURFUROL	COMP
HMG		HUMAN MENOPAUSAL GONADOTROPHEN	CLINCHEM
HMG	HMG-	3-HYDROXY-3-METHYLGLUTARYL-LIGAND	COMP
HMGR		HYDROXY METHYL GLUTAMYL COENZYM A REDUCTASE	CLINCHEM
HMI		HAHN-MEITNER INSTITUT (D)	INSTI
HMI		HEXAMETHYLENEIMINE	COMP
HMM		HEXAMETHYLMELAMINE	COMP
HMMA		4-HYDROXY-3-METHOXY MANDELIC ACID	COMP
HMN		2,2,4,4,46,8,8-HEPTAMETHYLNONANE	COMP
HMO		HUECKEL-MOLECULAR-ORBITAL-METHODE	THEOR
HMPA	HMPT	HEXAMETHYL-PHOSPHORIC ACID TRIAMIDE	COMP
HMPTA	HMPT HMPA	HEXAMETHYL-PHOSPHRICACID-TRIAMIDE	COMP
HMSA		HYDROXYMETHAN SULFONAT	COMP
HMSV		HARVEY MURINE SARCOMA VIRUS	CLINCHEM
HMT	HEXA HMTA	HEXANETHYLENE TETRAMINE	COMP POLYM
HMTT		3-HEXADECANOYL-4-METHOXYCARBONYL-1,3-THIAZOLIDINE-2-THIONE	COMP
HMV		HEART MINUTE VOLUME	CLINCHEM
HMV		HYDRODYNAMIC MODULATION VOLTAMMETRY	EANAL METH
HOBF		HIGHER ORDER BRIGHT FIELD	INSTR SPECT
HOBT		HYDROXYBENZTRIAZOLE	COMP
HODS		HIGHER ORDER DERIVATION SPECTROSCOPY	METH
HOG	PCP	1-(1-PHENYL CYCLOHEXYL) PIPERIDINE	COMP
HOL		HOLOGRAPHY	METH
HOMO		HIGHEST OCCUPIED MOLECULE ORBITAL	THEOR
HON		2-AMINO-5-HYDROXY-4-OXOPENTANOIC ACID	COMP

Abkürzung Akromym	Alternative	Bedeutung	Zuordnung
abbreviation akronym	alternative	meaning	related section
HONB		N-HYDROXY-5-NORBORNENE-2,3-DICARBOXYLIC ACID IMIDE	COMP
HOPG		HIGH ORIENTATED PYROGRAPHITE	MATER
HORSES		HIGHER ORDER RAMAN SPECTRAL EXCITATION STUDIES	METH
HOSA		HYDROXYLAMINE-O-SULFONIC ACID	COMP
HOSCH		HOCHLEISTUNGSSCHWEBSTOFF-FILTER	MATER
HP		HAEMATOPORPHYRIN	CLINCHEM
HP		HEWLETT PACKARD	COMPANY
HP		HIGH PRESSURE	TECHN
HP		HIGH PERFORMANCE	METH
HPA		HAEMATOPORPHYRIN ACETYLATED	CLINCHEM
HPA		2-HYDROXY PROPYL ACRYLATE	COMP
HPAA		HYDROXYPHENYL ACETIC ACID	COMP
HPBW		HALF PEAK BAND WIDTH	EVAL
HPC		HEALTH PROTECTION COMMITEE (CIFIC)	ORGANIS
HPDC	HPTLC	HIGH PERFORMANCE DUENNSCHICHTCHROMATOGRAPHIE	METH
HPETE	5-HPETE	5-HYDROPEROXY-6,8,11,14 EICOSATETRAENOIC ACID	COMP
HPHT		HIGH PRESSURE HIGH TEMPERATURE	TECHN
HPHT		HIGH PERFORMANCE LIQUIDCHROMATOGRAPHY WITH HYDROXYLAPATITE	METH
HPIC		HIGH PERFORMANCE ION CHROMATOGRAPHY	METH
HPIEC		HIGH PERFORMANCE ION EXCHANGE CHROMATOGRAPHY	METH
HPL		HUMAN PLACENTAL LACTOGEN	CLINCHEM
HPLAC		HIGH PERFORMANCE LIQUID AFFINITY CHROMATOGRAPHY	METH
HPLC		HIGH PERFORMANCE LIQUID CHROMATOGRAPHY	METH
HPLT	HPLTC	5-HYDROPEROXY LEUKOTRIENE (C)	COMP
HPMF		DIHYDRO-3,4-BIS((3-HYDROXYPHENYL)METHYL)-2(3H)-FURANONE	COMP
HPP		4-HYDROXY PYRAZOLO (3,4-d) PYRIMIDINE	COMP
HPPH		5-HYDROXYPHENYL-5-PHENYLHYDANTOIN	COMP
HPPLC		HIGH PERFORMANCE PREPARATIVE LIQUID CHROMATOGRAPHY	METH

Abkürzung Akromym	Alternative	Bedeutung	Zuordnung
abbreviation akronym	alternative	meaning	related section
HPPLC		HIGH PERFORMANE PLANAR LIQUID CHROMATOGRAPHY	METH
HPS		HEALTH PHYSICS SOCIETY (US)	ORGANIS
HPSEC		HIGH PERFORMANCE SICE EXCLUSION CHROMATOGRAPHY	METH
HPT		1-HYDROXY-2-PYRIDINTHION	COMP REAG
HPT		HIGH PERFORMANCE TITRIMETRY	METH
HPTLC	HPDC	HIGH PERFORMANCE THIN LAYER CHROMATOGRAPHY	METH
HPTS		1-HYDROXYPYREN-3.6.8-TRISULFONIC ACID	COMP
HPV		HUMAN PAPPILLOM VIRUS	CLINCHEM
HQNO		2-HEPTYL-4-QUINOLINOL 1-OXIDE	COMP
HR		HIGH RESOLUTION	METH
HRC		HUNTINGTON RESEARCH CENTRE (GB)	INSTI
HRE		HYPERE RAMAN EFFECT	METH SPECT
HRED		HIGH RESOLUTION ELCTRON DIFFRACTION	METH
HRELS		HIGH RESOLUTION ELECTRON LOSS SPECTROSCOPY	METH
HREM		HIGH RESOLUTION ELECTRON MICROSCOPY	METH
HRIR		HIGH RESOLUTION INFRARED RADIOMETRY	METH
HRMS		HIGH RESOLUTION MASS SPECTROMETRY	METH
HRP		HORSE RADISH PEROXIDASE	CLINCHEM
HRSGC		HIGH RESOLUTION SUBSTRACTION GAS CHROMATOGRAPHY	METH
HRSTEM		HIGH RESOLUTION SCANNING TRANSMISSION ELECTRON MICROSCOPY	METH
HRTEM		HIGH RESOLUTION TRANSMISSION MICROSCOPY	METH
HSA		HUMAN SERUM ALBUMIN	CLINCHEM
HSA		HEMISPERIC ANALYSER (DETECTOR)	SPECT
HSAB		HARD AND SOFT ACIDS AND BASES	THEOR
HSC	H.S.C.	BRITISH HEALTH AND SAFETY COMMISSION (GB)	INSTI
HSE	H.S.E.	HEALTH AND SAFETY COMMISSION EXCECUTIVE (GB)	INSTI
HSGC		HEADSPACE GAS CHROMATOGRAPHY	METH
HSIA		HALOGENATED SOLVENTS INDUSTRY ALLIANCE	ORGANIS
HSM		HERZSCHRITTMACHER	CLINCHEM

Abkürzung Akromym / abbreviation akronym	Alternative / alternative	Bedeutung / meaning	Zuordnung / related section
HSRI		HIGH SENSITIVITY REFRACTIVITY INDEX	EVAL
HSS		HEAD SPACE SAMPLER	INSTR
HSV		HERPES SIMPLEX VIRUS	CLINCHEM
HSW		HEALTH AND SAFETY WORK ACT (GB)	REG
HT	5-HT	5-HYDROXY TRYPTAMINE (=SEROTONINE)	COMP
HT		HIGH TEMPERATURE	TECHN
HTAB		HEXADECYLMETHYLAMMONIUMBROMIDE	COMP
HTG		HUMAN THRYOGLOBULIN	CLINCHEM
HTH		HYDRO-TOX-HYPOCHLORITE	CLINCHEM
HTL		HOEHERE TECHNISCHE LEHRANSTALT (CH)	INSTI
HTL		HIGH THRESHOLD LOGIC	ELECT
HTLV		HUMAN-T-CELL LEUKAEMIA VIRUS	CLINCHEM
HTMP		2,2,6,6-TETRAMETHYLPIPERIDINE	COMP
HTP	5HTP	5-HYDROXY TRYPTOPHAN	COMP
HTR		HIGH TEMPERATURE REACTOR	TECHN
HTS		HIGH TEMPERATURE SILYLATION	METH
HTS		HADAMAND-TRANSFER SPECTROMETRY	METH
HTSH	hTSH	HUMAN THYROID STIMULATING HORMONE	CLINCHEM
HTU		HIGHT OF TRANSFERUNIT (DETILLATION/GC)	METH
HTW		HOCHTEMPERATUR WINKLER VERFAHREN	TECHN
HV		HIGH VOLTAGE SUPPLIES	INSTR
HVA		HOMO VANILLIC ACID	COMP
HVCTEM		HIGH VOLTAGE CONVENTIONAL TRANSMISSION ELECTRON MICROSCOPY	METH
HVEM		HIGH VOLTAGE ELECTRON MICROSCOPY	METH
HVL		HYPOPHYSENVORDERLAPPEN HORMON	CLINCHEM
HVL		HALF VALUE LAYER (RADIOGRAPHY)	THEOR SPECT
HVSTEM		HIGH VOLTAGE SCANNING TRANSMISSION ELECTRON MICROSCOPY	METH
HWD	WLD,TCD	HOT WIRE DETECTOR	INSTR
HWM		HAZARDOUS WASTE MANGEMENT	ENVIR
HWZ		HALBWERTSZEIT	THEOR TECHN
HXIS		HARD X-RAY IMAGING SPECTROMETER	INSTR
HZO		HYDRATED ZIRKONIUMOXIDE	COMP

Abkürzung Akromym abbreviation akronym	Alternative alternative	Bedeutung meaning	Zuordnung related section
IAA	3-IAA,IES	INDOLE-3-ACETIC ACID	COMP
IAA		INSTITUT FÜR ARBEITSÖKONOMIK UND ARBEITSSCHUTZFORSCHUNG (DDR)	INSTI
IAE	IAe	INSTITUT FÜR AEROBIOLOGIE (FhG,D)	INSTI
IAEA		INTERNATIONAL ATOMIC ENERGY AGENCY (INT)	INSTI
IAEAC		INTERNATIONAL ASSOCIATION OF ENVIRONMENTAL ANALYTICAL CHEMISTRY (INT)	ORGANIS
IAEC		INTERNATIONAL ASSOCIATION OF ENVIRONMENTAL COORDINATORS (INT)	ORGANIS
IAEDANS	I-AEDANS	N-IODOACETYL-N'-(z-SULFO-NAPHTHYL) ETHYLENEDIAMINE (z=5 & 8)	COMP
IAEO	AIEA	INTERNATIONAL ATOMIC ENERGY ORGANIZATION (INT)	ORGANIS
IAES		INTERNATIONAL ACADEMY OF ENVIRONMENTAL SAFETY (INT)	CONF
IAES	I-AES	ION INDUCED AUGER SPECTROMETRY	METH
IAG		INTERN. ASSOCIATION FOR GERONTOLOGY (INT)	ORGANIS
IAG		INTERNATIONAL ACTION GROUP (CMA,CEFIC)	ORGANIS
IAHR		INTERNATIONAL ASSOCIATION FOR HYDRAULIC RESEARCH	ORGANIS
IAHS		INTERNATIONAL ASSOCIATION FOR HYDROLOGY SCIENCE	ORGANIS
IAM		INSTITUTE OF APPLIED MICROBIOLOGY UNIVERSITÄT TOKYO (JAP)	INSTI
IAMO		INTERNATIONAL ASSOCIATION OF MEDICAL OCEANOGRAPHY	ORGANIS
IAO	ILO	INTERNATIONALE ARBEITSORGANISATION (GENF,UN)	INSTI
IAOH		INTERNATIONAL ASSOCIATION FOR OCCUPATIONAL HEALTH	ORGANIS
IAPC		INSTITUT FÜR ANGEWANDTE PHYSIKALISCHE CHEMIE (KFA-JÜLICH,D)	INSTI
IAPS		INTER-AMERICAN PHOTOCHEMICAL SOCIETY	ORGANIS
IAPSO		INTERN. ASSOCIATION FOR PHYSICAL SCIENCES OF THE OCEAN (ISCU)	ORGANIS
IARC		INTERNATIONAL AGENCY FOR RESEARCH ON CANCER (INT)	INSTI

Abkürzung Akromym abbreviation akronym	Alternative alternative	Bedeutung meaning	Zuordnung related section
IASTED		INTERNATIONAL ASSOCIATION OF SCIENCE AND TECHNOLOGY FOR DEVELOPMENT (INT)	ORGANIS
IATA-DGR		INTERNATIONAL AIR TRANSPORT ASSOCIATION -DANGEROUS GOODS REGULATION INT)	REG
IATA-RAR		INTERNATIONAL AIR TRANSPORT ASSOCIATION-RESTRICTED ARTICLE REGULATION (INT)	REG
IAUR		INSTITUTIO DE ANTIBIOTICOS DA UNIVERSIDADE DE RECIFE (BRA)	INSTI
IAWPR		INTERNATIONAL ASSOCIATION ON WATER POLLUTION RESEARCH (INT)	ORGANIS
IAWR		INTERN. ARBEITSGEMEIN. DER WASSERWERKE IM RHEINEINZUGSGEBIET (EUR)	ORGANIS
IBA		4(3-INDOLYL) BUTYRIC ACID	COMP
IBAS		INTEGRALES BILD ANALYSEN SYSTEM	INSTR
IBC		INTERMEDIATE BULK CONTAINER	TECHN
IBD		IODOBENZENEDICHLORIDE	COMP
IBE		ION BEAM ETCHING	TECHN
IBG		INTERNATIONALE BIOMETRISCHE GESELLSCHAFT	ORGANIS
IBM		ISO-BUTYL METHYL ACRYLATE	COMP
IBMA		ISOBUTOXYMETHYL ACRYLAMIDE	COMP
IBMX		3-ISOBUTYL-1-METHYL XANTHINE	COMP
IBN		INSTITUTE BELGE DE NORMALISATION (BELG)	NORM
IBRG		INTERNATIONAL BIODEGRADATION GROUP	ORGANIS
IBSCA		ION BEAM SPECTROCHEMICAL ANALYSIS	METH
IBTMO		ISOBUTYL TRIMETHOXYSILANE	ICD
IC		ION CHROMATOGRAPHY	METH
IC	IC's	INTEGRATED CIRCUIT(S)	ELECT INSTR
IC		INITIAL CONDITIONS	EVAL
IC-ES		ION CHROMATOGRAPHY -ELUATION SUPPRESSION	METH
ICA		IGNITION CONTROL ADDITION	METH POLYM
ICAP		INDUCTIVELY COUPLED ARGON PLASMA	INSTR
ICASE		INTERNATIONAL COUNCIL OF ASSOCIATION OF SCIENCE EDUCATION (INT)	ORGANIS

Abkürzung Akromym abbreviation akronym	Alternative alternative	Bedeutung meaning	Zuordnung related section
ICC		INTERNATIONAL CHAMBER OF COMMERS (PARIS)	INSTI
ICCC		INTERNATIONAL CONFERENCE ON COORDINATION CHEMISTRY	CONF
ICCCRE		INTERN. CONFERENCE COMPUTERS IN CHEMICAL RESEARCH AND EDUCATION	CONF
ICCCS		INTERNATIONAL COMMITTEE ON CONTAMINATION CONTROL SOCIETIES	ORGANIS
ICD		ISOCITRIC DEHYDROGENASE	CLINCHEM
ICEAM		INTERNATIONAL COMMITEE OF ECONOMIC AND APPLIED MICROBIOLOGY (INT)	ORGANIS
ICERC		INTERNATIONAL CHEMICAL EMPLOYERS LABOR RELATIONS COMMITTEE	ORGANIS
ICES		INTERNATIONAL COUNCIL FOR THE EXPLORATION OF THE SEA (INT)	ORGANIS
ICF		INTRA CELLULARE FLUIDCONTENT	CLINCHEM
ICG		INDOCYANINE GREEN	COMP CLINCHEM
ICGEB		INTERNATIONAL CENTRE FOR GENETIC ENGINEERING AND BIOTECHNOLOGY (INT)	INSTI
ICGEC		INTERNATIONAL COLLABORATIVE GROUP ON ENVIRONMENTAL CARCINOGENESIS	ORGANIS
ICH		INSTITUT FÜR CHEMIE (KFA-JÜLICH,D)	INSTI
ICHEME	IChemE	INSTITUTE OF CHEMICAL ENGINEERS (GB)	INSTI
ICI		ISONAPHTHALOYLCHLORIDE	COMP
ICI		IMPERIAL CHEMICAL INDUSTRIES	COMPANY
ICIE		INTERNATIONAL CENTER FOR INDUSTRY AN ENVIRONMENT (UNEP,PARIS)	INSTI
ICIFI		INTERNATIONAL COUNCIL OF INFANT FOOD INDUSTRIES (INT)	ORGANIS
ICM		ION CONDUCTOR MODULATOR	CLINCHEM INSTR
ICOMC		INTERNATIONAL CONFERENCE ON ORGANOMETALLIC CHEMISTRY	CONF
ICOR		INTERNATIONAL CONFERENCE:OXYGEN REDICALS IN CHEMISTRY AND BIOLOGY	CONF
ICOS		INTERNATIONAL CONFERENCE ORGANIC SYNTHESIS (INT)	CONF
ICP		INDUCTIVELY COUPLED PLASMA	INSTR
ICP-OES		INDUCTIVELY COULED OPTICAL PLASMA EMISSION SPECTROMETRY	METH

Abkürzung Akromym abbreviation akronym	Alternative alternative	Bedeutung meaning	Zuordnung related section
ICP-SET		ICP-SAMPLE ELEVATOR TECHNIQUE	INSTR
ICPB		INTERNATIONAL COLLECTION OF PHYTOPATHOGENIC BACTERIA-UNIV. (DAVIS,US)	MICROB
ICPEMC		INTERN.COMMISSION FOR THE PROTECTION AGAINST ENVIRONMENTAL MUTAGENS AND CARCINOGENS	INSTI
ICPRAP		INTERNATIONAL COMMISSION FOR THE PROTECTION OF THE RHINE	INSTI
ICR		(PULSED) ION CYCLOTRON RESONANCE SPECTROSCOPY	METH
ICRDB		INTERNATIONAL CANCER RESEARCH DATA BANK	LIT
ICRP		INTERNATIONAL COMMISSION ON RADIOLOGICAL PROTECTION (INT)	ORGANIS
ICRU		INTERN. COMMISSION RADIATION UNITS (INT)	INSTI
ICS-OPTIC		INFINITY COLORCORRECTED SYSTEM	INSTR
ICSD		INORGANIC CRYSTAL STRUCTURE DATA	EVAL THEOR
ICSH		INTERSTITIAL CELL(ULAR) STIMULATING HORMONE	CLINCHEM
ICSPRO		INTER SECRETARIAT COMMITTEE ON SCIENTIFIC PROGRAMS RELATING TO THE OCEANOGRAPHY (UN)	INSTI
ICSU		INTERNATIONAL COUNCIL OF SCIENTIFIC UNIONS (INT)	ORGANIS
ICTA		INTERNATIONAL CONFEDERATION OF THERMAL ANALYSIS (INT)	ORGANIS
ICYP		3-IODOCYANOPINDALOL	COMP
ID		INHIBITORY DOSE	CLINCHEM
IDA	IDMS IVA	ISOTOPE DILUTION ANALYSES WITH MASS SPECTROMETRY	METH
IDA		INTELLIGENT DATA AQUISITION	EVAL ELECT
IDA		IMINO DIACETIC ACID	COMP
IDC		INTERN. DOKUMENTATIONSGESELLS. FÜR CHEMIE mbH (HAMBURG)	COMPANY
IDD		INSULIN DEPENDENT DIABETIS	CLINCHEM

Abkürzung Akromym abbreviation akronym	Alternative alternative	Bedeutung meaning	Zuordnung related section
IDEA		INTERNATIONAL CONGRESS ON DESALINATION AND WATER REUSE (INT)	CONF
IDHL		IMMEDIATELY DANGEROUS TO HEALTH OR LIFE	REG
IDLS		INTRACAVITY DYE LASER SPECTROMETRY	METH
IDMX		3-ISOBUTYL-1-METHXLXANTHINE	COMP
IDP		INOSINE-5'-DIPHOSPHATE	COMP CLINCHEM
IDP		IMAGE DEPTH PROFILING	METH
IDU	IDUR	5-IODO-2'-DEOXYURIDINE	COMP CLINCHEM
IE		ION EXCHANGE	METH
IE		IMMUNO ELECTROPHORESIS	METH
IEA		INTERNATIONAL ENERGY AGENCY	INSTI
IEB		INTERNATIONAL ENVIRONMENT BUREAU (INT)	INSTI
IEC		ION EXCLUSION CHROMATOGRAPHY	METH
IEC		INTERN. ELECTROTECHNICAL COMMISSION (INT)	INSTI
IEE	ESCA	INDUCED ELECTRON EMISSION	METH
IEE		INSTITUTION OF ELECTRICAL ENGINEERS (INT)	ORGANIS
IEF		ISO ELECTRIC FOCUSING	INSTR
IEHMO		ITERATIVE EXTENDED HUECKEL MOLECULAR ORBITAL	THEOR
IEMM		INCIDENT ENERGY MODULATION METHOD	METH
IENAA		INSTRUMENTAL ACTIVATION ANALYSIS WITH EPITHERMAL NEUTRONS	METH
IEP		IMMUNO ELECTROPHORESIS	METH
IEP		ISOELECTRIC POINT	EVAL
IEPA		INDEPENDENT ELECTRON PAIR APPROXIMATION	THEOR
IERL		INDUSTRIAL ENVIRONMENTAL RESEARCH LABORATORY (EPA,US)	INSTI
IES	IAA	INDOL-3-ESSIGSÄURE	COMP
IES		INSTITUT OF ENVIRONMENTAL SCIENCE	ORGANIS
IETAAS	I-ETAAS	IMPACTION-ELECTROTHERMAL ATOMIC ABSORPTION	METH
IETS		INELASTIC ELECTRON TUNNELING SPECTROSCOPY	METH
IEX	IE	ION EXCHANGE	INSTR
IEX		ION EXCITED X-RAY	INSTR

Abkürzung Akromym	Alternative	Bedeutung	Zuordnung
abbreviation akronym	alternative	meaning	related section
IEXC	IEC	ION EXCHANGE CHROMATOGRAPHY	METHJ
IEXF		IONEXCITED X-RAY FLUORESCENCE	INSTR
IEXS		ION EXCITED X-RAY SPECTROSCOPY	METH
IFA		IMMUNO FLUORESCENT ANTIBODY ASSAY	METH
IFA		INDUSTRIAL FRTILIZER INDUSTRY ASSOCIATION	ORGANIS
IFAG		INSTITUT FÜR ARZNEIMITTEL- UND GESUNDHEITSFORSCHUNG (DGU)	INSTI
IFAN		INETRNATIONALE FÖDERATION DER AUSSCHÜSS NORMENPRAXIS	ORGANIS
IFCB		INTERNATIONAL FEDERATION FOR CELL BIOLOGY (INT)	ORGANIS
IFCC		INTERNATIONAL FEDERATION CLINICAL CHEMISTRY (INT)	ORGANIS
IFD		IMMUNO FLUORESCENCE DETECTION	METH
IFE		IMMUNO FIXATION ELECTROPHORESIS	METH
IFEU		INSTITUT FÜR ENERGIE- UND UMWELTFORSCHUNG (HEIDELBERG)	INSTI
IFG		INSTITUT FÜR GÄRUNGSGEWERBE (BERLIN,D)	INSTI
IFM		INSTITUT FOR FOOD MICROBIOLOGY CHIBA-UNIVERS. (JAP)	INSTI
IFN	GAMMA-IFN	GAMMA INTERFERON	CLINCHEM
IFO		INSTITUT FOR FERMENTATION OSAKA (JAP)	INSTI
IFP		INCORPORATION FACTOR	CLINCHEM
IFP		INSTITUT FÜR ANGEWANDTE FESTKÖRPERPHYSIK (FhG,D)	INSTI
IFPMA	FIIM	INTERN.FEDERATION OF PHARMACEUTICAL MANUFACTURERS ASSOCIATION (INT)	ORGANIS
IFRA		INTERNATIONAL FRAGRANCE ASSOCIATION (INT)	ORGANIS
IFREP		INSTITUTE FRANCAIS DE RECHERCHES ET ESSAIS BIOLOGIQUE (F)	INSTI
IFS		INFRARED FOURIER SPECTROMETRY	METH
IFSCC		INTERNATIONAL FEDERATION OF SOCIETIES OF COSMETIC CHEMISTS (INT)	ORGANIS
IFSEM		INTERNATIONAL FEDERATION OF SOCIETIES FOR ELECTRON MICROSCOPY (INT)	ORGANIS
IFVTCC		INETRN.FÖDERATION DER VEREINE DER TEXILCHEMIKER UND COLORISTEN	ORGANIS

Abkürzung Akromym abbreviation akronym	Alternative alternative	Bedeutung meaning	Zuordnung related section
IG	Ig	IMMUNOGLOBULIN	CLINCHEM
IGA	IgA	IMMUNOGLOBULIN A	CLINCHEM
IGA		INTERESSENGEMEINSCHAFT AEROSOLE (FRANKFURT,D)	ORGANIS
IGC		INTERNATIONAL GASES COMMITTEE (PARIS)	ORGANIS
IGD	IgD	IMMUNOGLOBULIN D	CLINCHEM
IGEPHA		INTERESSENGEM. ÖSTERREICHISCHER HEILMITTELHERSTELLER (Ö)	ORGANIS
IGFET		INSULATED GATE FIELD EFFECT TRANSISTOR	INSTR
IGLO		INDIVIDUELL GAUGES FOR LOCATED ORBITAL	THEOR
IGR		INSECT GROWTH REGULATOR	BIOS
IGSS		IMMUNO GOLD SILVER STAINING	CLINCHEM
IGU		INTERNATIONAL GAS UNION (INT)	ORGANIS
IH		IMMOBILIZED HISTAMINE	CLINCHEM
IHBM		INSTITUT FÜR HYGIENE DER BUNDESANSTALT FÜR MILCHFORSCHUNG (KIEL,D)	INSTI
IHK		INDUSTRIE UND HANDELSKAMMER (D)	INSTI
IHM		INSTITUTO DE HIGIENE EXPERIMENTAL MOTEVIDEO (URU)	INSTI
IHME		INTERNATIONAL HOSPITAL MEDICAL EXIBITION (INT)	CONF
IIASA		INTERNATIONALES INSTITUT FÜR ANGEWANDTE SYSTEMANALYSE (ÖSTER)	INSTI
IICCI		INETRNATIONAL INFORMATION CENTRE OF COSMETICS INDUSTRIES	ORGANIS
IICT		INETRNATIONAL INSTITUT OF CHEMICAL TOXICOLOGY	INSTI
IID		ION IMPACT DESORPTION	INSTR
IID		ION INDUCED DESORPTION	METH
IIDQ		2-ISOBUTOXY-1-ISOBUTOXYCARBONYL-1,2-DIHYDROQUINOLINE	COMP
III		INTERNATIONAL ISOCYANATE INSTUTE (NEW YORK)	INSTI
IILE		ION INDUCED LIGHT EMISSION	METH
IIR		ISOBUTYLENE-ISOPRENE-RUBBER	MATER POLYM
IIR		ION INDUCED RADIATION =X-RAY EMISSION INDUCED BY ION BOMBARDMENT	METH
IIRS		ION INDUCED RADIATION SPECTROSCOPY	METH

Abkürzung Akromym	Alternative	Bedeutung	Zuordnung
abbreviation akronym	alternative	meaning	related section
IISRP		INTERNATIONAL INSTITUT OF SYNTHETIC RUBBER PRODUCERS	INSTI
IITR		ILLINOIS INSTITUTE OF TECHNOLOGY RESEARCH (US)	INSTI
IIUG		INTERNATIONALES INSTITUT FÜR UMWELT UND GESELLSCHAFT (BERLIN,D)	INSTI
IIXE		ION INDUCED X-RAY EMISSION	INSTR
IIXS		ION INDUCED X-RAY SPECTROSCOPY	METH
IJFM		INSTITUTO JAIME FERRAN DE MICROBIOLOGIA (MADRID,ESP)	INSTI
IKES		ION KINETIC ENERGY SPECTROSCOPY	METH
IKS	OICM	INTERKANTONALE KONTROLLSTELLE FUER HEILMITTEL (CH)	INSTI
IKSR	ICPRAP	INTERNATIONALE KOMMISSION ZUM SCHUTZE DES RHEINS	INSTI
IKV		INSTITUT FUER KUNSTSTOFF VERARBEITUNG IN INDUSTRIE UND HANDWERK (D)	ORGANIS
IKV		INTERKANTONALE VEREINBARUNG (CH)	REG
IKW		INDUSTRIEVERBAND KÖRPERPFLEGE UND WASCHMITTEL (D)	ORGANIS
ILE	Ile	ISOLEUCINE	COMP
ILEED		INELASTIC LOW ENERGY ELECTRON DIFFRACTION	METH
ILIB		INTERNATIONAL LIPID INFORMATION BUREAU	LIT
ILL		INSTITUTE LAUE LANGVIN GRENOBLE (CH,EUR)	INSTI
ILMAC		INTERN.CHEMIEFACHMESSE FÜR LABORATORIUMS-,VERFAHRENS-,MESSTECHNIK UND AUTOMATION (CH)	CONF
ILO		INTERNATIONAL LABOR ORGANIZATION (UN)	ORGANIS
ILS		IONIZATION LOSS SPECTROSCOPY	METH
ILZRO		INTERNATIONAL LEAD AND ZINC RESEARCH ORGANIZATION	ORGANIS
IM	i.m.	INTRA MUSCULAER	CLINCHEM
IM		IMAGE MATCHING	METH
IMCO		INTERGOVERNMENTAL MARITIME CONSULTATIVE ORGANIZATION (INT)	ORGANIS
IMDG		INTERNATIONAL MARITIME DANGEROUS GOODS CODE (INT)	REG

Abkürzung Akromym abbreviation akronym	Alternative alternative	Bedeutung meaning	Zuordnung related section
IMEO		IMIDAZOLINEPROPYLTRIETHOXYSILANE	COMP
IMER		IMMOBILIZED ENZYME REACTOR	INSTR
IMET		ZENTRALSTELLE FÜR MIKROBIOLOGIE UND EXPERIMENTELLE THERAPIE (JENA,DDR)	INSTI
IMG		INSTITUT FÜR MIKROBIOLOGIE DER UNIVERSITÄT GÖTTINGEN (D)	INSTI
IMGE		IONSELECTIVE MULTISOLVENT GRADIENT ELUTION	METH
IMMA		ION MICROPROBE MASS ANALYSIS	METH
IMO		INTERNATIONAL MARITIME ORGANIZATION (INT)	REG
IMP		INOSINE-5'-MONOPHOSPHATE	COMP CLINCHEM
IMP		ION MODERATED PARTITION	METH
IMP		N-ISOPROPYL-4-IODOAMPHETAMINE	CLINCHEM COMP
IMP		ISOCRATIC MULTISOLVENT PROGRAMMING	INSTR
IMPY		2,3-DIHYDRO-1H-IMIDAZOL(1,2-b)PYRAZOLE	COMP REAG
IMS		ISOCRATIC MULTISOLVENT	INSTR
IMXA		ION MICROPROBE X-RAY ANALYSIS	METH
INAA		INSTRUMENTAL NEUTRON ACTIVATION ANALYSIS	METH
INADEQUATE		INCEDIBLE NATURAL ABUNDANCE SAMPLE QUANTUM TRANSFER EXPERIMENT (NMR)	METH
INAH	INH	ISONICOTINIC ACID HYDRAZIDE	COMP
IND		INDENO (1,2,3 cd)PYREN	COMP
INDO		INTERMEDIATE NEGLECT OF DIFFERENTIAL OVERLAP	THEOR
INDOR		INTERMOLECULAR DOUBLE RESONANCE	METH
INEPT		INSENSITIVE NUCLEI ENHANCEMENT BY POLARIZATION TRANSFER (NMR)	METH
INFU		INSTITUT FUER UMWELTSCHUTZ (UNI DORTMUND,D)	INSTI
INFUCHS		INFORMATIONSSYSTEM FÜR UMWELTCHEMIKALIEN UND STOFFE (UBA,D)	LIT
INH	INAH	ISONICOTINIC ACID HYDRAZIDE =ISONIAZID	COMP
INHG		GLYCONIAZID	CLINCHEM COMP
INIS		INTERNATIONAL NUCLEAR INFORMATION SERVICE	LIT
INMS		IONIZED NEUTRAL MASS SPECTROMETRY	METH
INN		INTERNATIONAL NON-PROPRIETY NAME	COMP REG

| Abkürzung Akromym | Alternative | Bedeutung | Zuordnung |
abbreviation akronym	alternative	meaning	related section
INPADOC		INTERNATIONALES PATENT DOKUMENTATIONSZENTRUM (INT)	INSTI
INRS		INSTITUTE NATIONAL DE RECHERCHE ET DE SECURITE (F)	INSTI
INS		6((4-METHYLPHENYL)AMINO)-2-NAPHTHALIN SOLFONATE	COMP
INS		ION NEUTRALIZATION SPECTROSCOPY	METH
INT		2-(4-IODOPHENYL)-3-(4-NITROPHENYL)-5-PHENYLTETRAZOLIUM CHLORIDE	COMP
IOBC		INTERNATIONAL ORGANIZATION FOR BIOLOGICAL CONTROL	ORGANIS
IOBS		INTEGRATED OPTICAL BIOSENSOR	INSTR
IOC		INTERNATIONAL OCEAN COUNCIL OF THE UNESCO (INT)	INSTI
IOCD		THE INTERN. ORGANIZATION FOR CHEMICAL SCIENCES IN DEVELOPMENT (INT)	ORGANIS
IOFI		INTERNATIONAL ORGANIZATION OF THE FLAVOUR INDUSTRY (INT)	ORGANIS
IOI		INTERNATIONAL OZON INSTITUT	INSTI
IP	I.P.	THE PHARMACOPOEIEA OF INDIA	NORM
IP		ION PROBE	INSTR
IP		ISOPHORONE (=3,5,5-TRIMETHYL-2-CYCLOHEXANONE)	COMP REAG
IP3		INOSITOL TRIPHOSPHATE	COMP
IPA		ISOPROPYLALKOHOL	COMP
IPA		INTERNATIONAL PLASTIC ASSOCCIATION	ORGANIS
IPAA		1-HYDROXY-3-AMINOMETHYL-3,5,5-TRIMETHYLCYCLOHEXANE	COMP
IPC		ISOPROPYL-N-PHENYLCARBAMATE	COMP
IPC		ION PAIR CHROMATOGRAPHY	METH
IPC		INDIRECT PHOTOMETRIC CHROMATOGRAPHY	METH
IPCR		INSTITUT FOR PHSICAL AND CHEMICAL RESEARCH (SAITAMA,JAP)	INSTI
IPCS		INTERNATIONAL PROGRAM ON CHEMICAL SAFETY (UN)	TECHN
IPD		1-AMINO-3-AMINOMETHYL-3,5,5-TRIMETHYLCYCLOHEXANE	COMP
IPDI		ISOPHORONE DIISOCYANATE (=3-ISOCYANATOMETHYL-3,5,5-TRIMETHYLCYCLOHEXYLISOCYANATE)	COMP

Abkürzung Akromym	Alternative	Bedeutung	Zuordnung
abbreviation akronym	alternative	meaning	related section
IPG	IPG's	IMMOBILIZED ph GRADIENTS	CLINCHEM
IPIECA		INTERN.PETROLEUM INDUSTRY ENVIRONMENTAL CONSERVATION ASSOCIATION	ORGANIS
IPM		INTEGRATED PEST MANAGEMENT (AGRICULTURE)	TECHN
IPN		ISOPHTHALONITRILE	COMP
IPN	IPN's	INTERPENETRATING POLYMER NETWORKS	MATER POLYM
IPOTMS		ISOPROPENYLOXYTRIMETHYL SILANE	COMP
IPP		ION PAIR PARTITION	METH
IPP		MAX PLANCK INSTITUT FUER PLASMAPHYSIK GARCHING (D)	INSTI
IPPB		INTERMITTING POSITIVE PRESSURE BREATHING	CLINCHEM
IPPC		ISOPROPYL-PHENYL-CARBAMATE	MATER
IPS		INDUSTRIEVERBAND PFLANZENSCHUTZ	ORGANIS
IPSD		INTEGRATING POSITION SENSITIVE DETECTOR	INSTR
IPTG		ISOPROPYL BETA-D-THIOGALACTOSIDE	COMP CLINCHEM
IPV		INSTITUTO DE PATOLOGIA VEGETALE UNIVERSITA DI MILANO (I)	INSTI
IR		INFRARED	INSTR
IR		INDUCED RADIOACTIVITY	INSTR
IR		ISOPRENE RUBBER	MATER POLYM
IRANOR		INSTITUTO NATIONAL DE RACIONALIZACON Y NORMALIZACON (SP)	INSTI
IRC		INTERNATIONALER RECYLING CONGRESS	
IRCD		INFRARED CIRCULAR DICHROISM	METH
IRCHA		INSTITUTE NATIONAL DE RECHERCHE DE LA CHIMIE APPLICQUEE (F)	INSTI
IRD		INFRARED DETECTOR	INSTR
IRE		INTERNAL REFLECTION ELEMENTS	INSTR
IRLG		INTERAGENCY REGULATORY LIAISON GROUP (US)	INSTI
IRM		INTERIM REFERENCE MATERIAL	MATER METH
IRMA		IMMUNO RADIOMETRICAL ASSAY	METH
IRMA		IMMISSIONSRATEN MESSAPPARATUR (VDI,D)	INSTR
IRMC		INTERAGENCY RISK MANGEMENT COUNCIL (US)	INSTI

Abkürzung Akromym abbreviation akronym	Alternative alternative	Bedeutung meaning	Zuordnung related section
IRMPD		INFRARED MULTIPLE-PHOTON DECOMPOSITION	INSTR
IRP		INFRARED PYROMETRY	METH
IRPA		INTERN. RADIATION PROTECTION AGENCY (INT)	INSTI
IRPA		INTERNATIONAL RADIATION PROTECTION ASSOCIATION	ORGANIS
IRPTC		INTERNATIONAL REGISTER OF POTENTIALLY TOXIC CHEMICALS	LIT
IRRAS	IR-RAS	INFRARED REFLEXION ABSORPTION SPECTROSCOPY	METH
IRRS	IRS	INFRARED REFLECTANCE SPECTROSCOPY	METH
IRS		INFRARED SPECTROSCOPY	METH
IRS		INVERSE RAMAN-SPECTROSCOPY	METH
IRS		INTERNAL REFLECTANCE SPECTROMETRY	METH
IRS		INTERNATIONAL ENVIRONMENT REFERENCE SYSTEM (UN)	LIT
IS		IONIZATION SPECTROSCOPY	METH
ISA		INTERNAL STANDARD ADDITION	EVAL METH
ISAS		INSTITUT FUER SPECTROCHEMIE UND ANGEWANDTE SPECTROSCOPIE (DORTM,D)	INSTI
ISBN		INTERNATIONAL STANDARD BOOK NUMBER (LIT)	LIT
ISC		INTERSYSTEM CROSSING (NMR)	INSTR
ISC		INTERNATIONAL SOCIETY OF CHEMOTHERAPY	ORGANIS
ISCC		INTERNATIONAL SYMPOSIUM ON CAPILLARY CHROMATOGRAPHY (INT)	CONF
ISCC		INTER-SOCIETY COLOUR COUNCIL	ORGANIS
ISCE		INTERNATIONAL SOCIETY OF CLINICAL ENZYMOLOGY (INT)	ORGANIS
ISCOM	ISCOMS	IMMUNO STIMULATED COMPLEXES	CLINCHEM
ISD		ION STIMULATED DESORPTION	INSTR
ISDN		ISOSORBIDE DINITRATE	COMP
ISE		ION SELECTIVE ELECTRODE	INSTR
ISE		INTERNATIONAL SOCIETY OF ELECTROCHEMISTRY (INT)	ORGANIS
ISEES		INTERNATIONAL SOCIETY FOR ECOTOXIXOLGY AND ENVIRONMENTAL SAFETY	ORGANIS

Abkürzung Akromym	Alternative	Bedeutung	Zuordnung
abbreviation akronym	alternative	meaning	related section
ISERC		INSTITUT SUISSE DE RECHERCHES EXPERIMENTALES SUR LE CANCER (CH)	INSTI
ISF		INTERNATIONAL SOCIETY FOR FATRESEARCH (INT)	ORGANIS
ISFET		ION SENSITIVE FIELD EFFECT TRANSISTOR	INSTR
ISHC		INTERNATIONAL SOCIETY OF HETEROCYCLIC CHEMISTRY (INT)	ORGANIS
ISHC		INTERNATIONAL SYMPOSIUM ON HOMOGENOUS CATALYSIS	CONF
ISI		INSTITUTE FOR SCIENCE INFORMATION (US)	LIT
ISMA		INTERN. SUPERPHOSPHATE AND COMPOUND MANUFATURERS ASSOCIATION	ORGANIS
ISO		INTERNATIONAL STANDARDS ORGANIZATION (INT)	ORGANIS NORM
ISP		INTERNATIONAL STREPTOMYCES PROJECT WESLEYAN UNIVERS. (DELEWARE,US)	MICROB
ISRP		INTERNAL SURFACE REVERSE PHASE	MATER
ISS		IONENSPEKTROSKOPIE	METH
ISS	LEIS	ION SCATTERING SPECTROSCOPY	METH
ISSN		INTERNATIONAL STANDARD SERIAL NUMBER	LIT
ISTEM		INTERNAL SCANNING TRANSMISSION ELECTRON MICOROSCOPY	METH
ITA		ITACONIC ANHYDRIDE	COMP
ITC		INTERNATIONAL TRIBOLOGY COUNCIL (LONDON)	ORGANIS
ITC		INTERAGENCY TESTING COMMITTEE (US)	ORGANIS
ITCC		INDIAN TYPE CULTURE COLLECTION (NEW DEHLI,IND)	INSTI
ITD		ION TRAP DETECTOR (MASS SPECTROSCOPY)	INSTR
ITNAA		INSTRUMENTAL ACTIVATION ANALYSIS WITH THERMAL NEUTRONS	METH
ITO		INDIUM TRI OXIDE	COMP
ITP		INOSINE-5' TRIPHOSPHATE	COMP
ITP		ISO-TACHO-PHORESE	METH
ITS		INELASTIC TUNNELEING SPECTROSCOPY	METH
ITSDC		INTERAGENCY TOXIC SUBSTANCES DATA COMMITTEE (US)	INSTI
IU	I.U.	INTERNATIONAL UNIT	EVAL

Abkürzung Akromym abbreviation akronym	Alternative alternative	Bedeutung meaning	Zuordnung related section
IUAPPA	IUPPA	INTERN. UNION OF AIR POLLUTION PREVENTION ASSOCIATION	ORGANIS
IUB		INTERNATIONAL UNION OF BIOCHEMISTRY (INT)	ORGANIS
IUC	I.U.C.	INTERNATIONAL UNION OF CHEMISTRY (INT)	INSTI
IUCN		INTERN. UNION FOR CONSERVATION OF NATURE AND NATURAL RESOURCES (UN)	INSTI
IUFOST		INTERNATIONAL UNION OF FOOD SCIENCE AND TECHNOLOGY	ORGANIS
IUMS		INTERNATIONAL UNION OF MICROBIOLOGICAL SOCIETIES (INT)	ORGANIS
IUNS		INTERNATIONAL UNION AT NUTRITIONAL SCIENCES (INT)	ORGANIS
IUPAB		INTERNATIONAL UNION OF PURE AND APPLIED BIOPHSICS (INT)	INSTI
IUPAC		INTERNATIONAL UNION OF PURE AND APPLIED CHEMISTRY (INT)	ORGANIS NORM
IUPAP		INTERNATIONAL UNION OF PURE AND APPLIED PHYSICS (INT)	ORGANIS
IUPHAR	IUPhar	INTERNATIONAL UNION OF PHRMACOLOGY (INT)	ORGANIS
IUTOX		INTERNATIONAL UNION OF TOXICOLOGY	ORGANIS
IUVSTA		INTERNATIONAL UNION FOR VACUUM SCIENCE TECHNIQUES AND APPLICATION (INT)	ORGANIS
IV	i.v.	INTRAVENOES	CLINCHEM
IVA	IDA	ISOTOPENVERDUENNUNGS ANALYSE	METH
IVC		INDUSTRIEVEREINIGUNG CHEMIEFASERN	ORGANIS
IVH		INDUSTRIEVERBAND HARTSCHAUM e.V. (D)	ORGANIS
IVI		IN VITRO INTERNATIONAL INC. (ANN ARBOR,US)	COMPANY
IVPU		INDUSTRIEVERBAND POLYURETHAN HARTSCHAUM e.V. (D)	ORGANIS
IWL		INSTITUT FÜR GEWERBL.WASSERWIRTSCHAFT UND LUFTREINHALTUNG e.V.(KÖLN)	INSTI
IWSA		INTERNATIONAL WATER SUPPLY ASSOCIATION	ORGANIS
IZAA		5-CHLOROINDAZOL-3-ACETIC ACID ETHYLESTER	COMP

Abkürzung Akromym	Alternative	Bedeutung	Zuordnung
abbreviation akronym	alternative	meaning	related section
JACC		JOINT ASSESSMENT OF COMMODITY CHEMICAL	ORGANIS
JACR		JAPAN AIR CLEANING ASSOCIATION (J)	ORGANIS
JAIC		JAPAN ASSOCIATION FOR INTERN. CHEMICAL INFORMATION (J)	ORGANIS
JAZ		JAHRES AKTIVITAETS ZUFUHR	REG TECHN
JCIA		JAPAN COSMETIC INDUSTRY ASSOCIATION (JAP)	ORGANIS
JECFA		JOINT EXPERT COMMITTEE ON FOOD ADDITIVES (FAO-WHO)	INSTI
JET		JOINT EUROPEAN TOROUS (NUCLEAR FISSION PLANT)	TECHN
JETOC		JAPANESE CHEMICAL INDUSTRY ECOLOGY & TOXICOLOGY CENTRE (JAP)	INSTI
JIS		JAPANESE INDUSTRIAL STANDARDS (J)	NORM
JOCE		JOURNAL OFFICE DES CE (BRÜSSEL)	LIT
JP	J.P.	THE PHARMACOPOEIA OF JAPAN (J)	NORM
JPA		JAPANESE PHOTOCHEMISTRY ASSOCIATION (JAP)	ORGANIS
JRC		JOINT RESEARCH CENTRE OF THE CE (EG)	INSTI
JSDA		JAPAN SOAP AND DETERGENT ASSOCIATION (JAP)	ORGANIS
JTE		JAHN-TELLER EFFECT	THEOR
JUNIC		JOINT UNITED NATIONS INFORMATION COMMITTEE (UN)	LIT

Abkürzung Akromym	Alternative	Bedeutung	Zuordnung
abbreviation akronym	alternative	meaning	related section
KAM		KALMOGOROW, ARNOLD AND MOSER CRITERIA	THEOR
KAP		KALIUMHYDROGENPHTHALATE	COMP
KAPA		KALIUM 3-AMINOPROPYLAMID	COMP
KBA		3-KETOBUTYRALDEHYDE DIMETHYLACETAL	COMP
KBT		4-KETOBENZTRIAZINE	COMP
KDO		2-KETO-3-DIOXYOCTANATE	COMP
KDT		KAMMER DER TECHNIK (DDR)	INSTI
KDV		KOHLE DRUCK VORGANG	TECHN
KEIDANREN		FEDERATION OF JAPANESE INDUSTRY (TOKYO,JAP)	ORGANIS
KEWA		KERNBRENNSTOFF-WIEDERAUFBEREITUNGS-GESELLSCHAFT mbH (FRANKFURT,D)	COMPANY
KFA		KERNFORSCHUNGS ANLAGE JUELICH (D)	INSTI
KFAA		KURATORIUN FÜR DAS FORTBILDUNGSZENTRUM ABWASSER UND ABFALL (ESSES,D)	ORGANIS
KFK	KfK	KERNFORSCHUNGSZENTRUM KARLSRUHE (D)	INSTI
KFLSG	K-F-Lsg	KARL FISCHER LOESUNG	MATER
KFT		KARL FISCHER TITRATION	METH
KFW	KfW	KURATORIUM FÜR WASSERWIRTSCHAFT (D)	ORGANIS
KHP		KALIUM HYDROGEN PHTHALAT	COMP
KHTAB		8-KETOHEXADECYL TRIMETHYLAMMONIUM BROMIDE	COMP
KIB		ARBEITSKREIS SELBSTÄNDIGER KUNSTSTOFF-INGENIEURE UND BERATER (D)	ORGANIS
KIPBH	KIP-BH	KALIUM TRIISOPROPOXY BORHYDRID	COMP
KISS		KOSSEL INTERNAL STRESS METHOD	THEOR
KITA		KITASATO INSTITUTE TOKYO DEPARTMENT OF ANTIBIOTICS (JAP)	INSTI
KIWA		KEURINGSINSTITUT VOOR WATERLEIDINGARTIKELEN (NDL)	INSTI
KK		KEMIAN KESKULILITTO (HELSINKI,FIN)	INSTI
KKKK		KALLIKREIN-KINEN-KININOGEN KINASE	CLINCHEM
KKN		KERNKRAFTWERK NIEDERAICHACH (D)	INSTI
KKW		KERNKRAFTWERK	TECHN
KKZ		KONDENSATIONS KEIM ZÄHLER	TECHN

Abkürzung Akromym	Alternative	Bedeutung	Zuordnung
abbreviation akronym	alternative	meaning	related section
KLF		KIEMIKALIE LEVERANDORENES FORENING (NOR)	ORGANIS
KLH		KEYHOLE LIMPET HAEMOCYANINE	CLINCHEM
KLM		KOMMISSION LÄRMMINDERUNG (VDI,D)	ORGANIS
KM		KAMAMYCIN	CLINCHEM
KMDP		KOSSEL-MOELLENSTADT DIFFRACTION PATTERN	THEOR
KNCV		KONINKLIJKE NEDERLANDSE CHEMISCHE VERENIGING (NLD)	ORGANIS
KNMP		KONINKLIJK NEDERLANDSE MAATSCHAPPIG TER BEVORDERING PHARMACIE (NDL)	ORGANIS
KS		KAMMERSAETTIGUNG (ANGABE BEI DER DC)	INSTR
KSAM		ARBEITSKR.KRISTALLSTRUKTURANALYSE VON MOLEKUELVERB. (GDCh,D)	ORGANIS
KSR	NMR	KERNSPINRESONANZ	METH
KST		KERNSPINTOMOGRAPH	INSTR
KTA		KERNTECHNISCHER AUSSCHUß (BMI,D)	ORGANIS
KTG		KERNTECHNISCHE GESELLSCHAFT (D)	ORGANIS
KTÖ		KOORDINIERUNGSAUSSCHUß TOXIKOLOGIE/ ÖKOLOGIE (VCI	ORGANIS
KUER		KOMMISSION FÜR DIE ÜBERWACHUNG DER READIOAKTIVITÄT	INSTI
KWF		KOMMISSION ZUR FOERDERUNG DER WISSENSCHAFTLICHEN FORCHUNG (CH)	INSTI

Abkürzung Akromym / abbreviation akronym	Alternative / alternative	Bedeutung / meaning	Zuordnung / related section
LA		LATEX AGGLUTINATION	CLINCHEM
LAAO		L-AMINO ACID OXIDASE	CLINCHEM
LAC		LIQUID ADSORPTION CHROMATOGRAPHY	METH
LACBED		LARGE ANGLE COVERGENT BEAM ELECTRON DIFFRACTION	METH
LADB		LABORATORY ANIMAL DATA BANK	LIT
LADH		LIVER ACOHOL DEHYDROGENASE	CLINCHEM
LAF		LASER ATOMIC FLUORESCENCE SPECTROSCOPY	METH
LAGA		LANDES ARBEITSGEMEINSCHAFT ABFALL (D)	INSTI
LAH		LITHIUM ALUMINIUM HYDRIDE	COMP
LAI		LATEX AGGLUTINATION INHIBITION	CLINCHEM
LAI		LÄNDERAUSSCHUß IMMISSIONSSCHUTZ (D)	INSTI
LAK		LANDESAPOTHEKERKAMMER (D)	INSTI
LAL		LIMULUS-AMÖBOZYTEN LYSAT-TEST	CLINCHEM
LALLS		LOW-ANGLE LASER LIGHT SCATTERING	METH
LAMES		LASER MICRO EMISSION SPECTROSCOPY	METH
LAMMA		LASER MICROPROBE MASS ANALYSIS	METH
LAMMS		LASER MICRO MASS SPECTROMETRY	METH
LAN		LOCAL AREA NETWORK	ELECT TECHN
LAP		LEUCINE ARYLAMIDASE PEPTIDHYDROLASE	CLINCHEM
LARIS		LASER ABLATION AND RESONANCE IONIZATION SPECTROMETRY	METH
LAS		LABORATORY AUTOMATION SYSTEM	TECHN
LASER		LIGHT AMPLIFICATION BY STIMULATED EMISSION OF RADIATION	INSTR
LATGS	L-ATGS	L-ALANINE ENDOWED TRIGLYCINESULFATE (IR DETECTION)	MATER
LAV		LYMPHADINOPATHY ASSOCIATED VIRUS	CLINCHEM
LAV		LANDESAPOTHEKERVEREIN (D)	ORGANIS
LAW		LOW ACTIVE WASTE	TECHN
LAWA		LAENDER ARBEITSGEMEINSCHAFT WASSER UND ABFALL (D)	INSTI
LBF		LABORATORIUM FÜR BETRIEBSFESTIGKEIT (FhG,D)	INSTI
LBK		LECTURE BOTTLE FOR CORRESSIVE GAS	MATER TECHN
LBN		LECTURE BOTTLE FOR NON CORRESSIVE GAS	MATER TECHN
LC		LIQUID CHROMATOGRAPHY	METH
LCAO		LINEAR COMBINATION OF ATOMIC ORBITALS	THEOR

Abkürzung Akromym abbreviation akronym	Alternative alternative	Bedeutung meaning	Zuordnung related section
LCBO		LINEAR COMBINATION OF BOND-ORBITALS	THEOR
LCCC		LOCULAR COUNTER CURRENT CHROMATOGRAPHY	METH
LCD		LIQUID CRYSTAL DIODES & LIQUID CRYSTAL DISPLAY	INSTR
LCEC	LCECD	LIQUID CHROMATOGRAPHY WITH ELECTROCHEMICAL DETECTION	METH
LCF	LWL	LIGHT CONDUCTING FIBRE	MATER
LCFC		LINEAR COMBINATION OF FRAGMENT CONFIGURATION	THEOR
LCMO		LINEAR COMBINATION OF MOLECULAR ORBITALS	THEOR
LCP		LIQUID CRYSTAL POLYMER	MATER
LCP		LABORATOIRE DE CRYPTOGAME MUSEE NATION.D'HISTOIRE NATUREL PARIS (F)	INSTI
LCR		LEUROCRISTINE	CLINCHEM
LCUV		LIQUID CHROMATOGRAPHY WITH UV DETECTION	METH
LD		LETALE DOSIS	CLINCHEM
LD		LASER DESORPTION	INSTR
LDA		LASER DOPPLER ANEMOMETRY	METH
LDA		LITHIUM DIISOPROPYLAMIDE	COMP
LDA		LINEAR DIODE ARRAY	INSTR
LDAO		N,N-DIMETHYL DODECYLAMINE-N-OXIDE	COMP
LDF		LOKALER DICHTE FORMALISMUS	THEOR
LDH		LACTIC DEHYDROGENASE	CLINCHEM
LDL		LOW DENSITY LIPOPROTEIN	CLINCHEM
LDLC		LOW DISPERSION LIQUID CHROMATOGRAPHY	METH
LDMAN		LITHIUM DIMETHYLAMINONAPHTHALENE	COMP
LDME		LASER DOPPLER MICROELECTROPHORESIS	METH
LDPE		LOW DENSITY POLYETHYLENE	MATER POLYM
LDV		LASER DOPPLER VELOCIMETRY	METH
LE		LUPUS ERYTHREMATHODES	CLINCHEM
LEAC		LINEAR ELUTION ADSORPTION CHROMATOGRAPHY	METH
LEAFS		LASER EXCITED ATOMIC FLUORESCENCE SPECTROMETRY	METH
LEC	LEC-VERF	LIQUID-ENCAPSULATED CZOCHRALSKI-VERFAHREN	TECHN

Abkürzung Akromym	Alternative	Bedeutung	Zuordnung
abbreviation akronym	alternative	meaning	related section
LED	LED's	LIGHT EMITTING /LUMINESCENCE DIODE(S)	INSTR
LEED		LOW ENERGY ELECTRON DIFFRACTION	METH
LEEDOA		LOW ENERGY ELECTRON DIFFRACTION OPTICS ANALYZER	INSTR
LEEIXS		LOW ENERGY ELECTRON INDUCED X-RAY SPECTROSCOPY	METH
LEELS		LOW ENERGY ELECTRON LOSS SPECTROSCOPY	METH
LEEM		LOW ENERGY ELECTRON MIROSCOPY	METH
LEES		LOW ENERGY ELECTRON SPECTROMETRY	METH
LEI		LASER ENHANCHED IONIZATION	METH
LEIS		LOW ENERGY ION SCATTERING SPECTROSCOPY	METH
LEM		LORENTZ ELECTRON MICROSCOPY	METH
LEM		LEUCOCYTE ENDOGENOUS MEDIATOR	CLINCHEM
LEMBS		LOW ENERGY MOLECULAR BEAM SCATTERING	METH
LEMO		LOWEST EMPTY MOLECULAR ORBITAL	THEOR
LEPS		LOW ENERGY PHOTON SPECTROSCOPY	METH
LERD		LOW ENERGY X-RAY DIFFRACTION	METH
LES		LICENSING EXCECUTIVE SOCIETY (INT)	ORGANIS
LESS		LASER EXCITED SHPOL'SKII-SPECTROMETER	INSTR
LET		LOW ENERGY TEMPERATURE	THEOR
LET		LINEAR ENERGY TRANSFER	TECHN
LEU	Leu	LEUCINE	COMP
LEUPA		L-LEUCIN-4-NITROANILID	COMP CLINCHEM
LEX		LOW ENERGY X-RAY SPECTRAL ANALYSIS	METH
LF	LFC	LAMINAR FLOW (CABINET)	TECHN
LFL		LOWER FLAMMABLE LIMIT	TECHN
LFU	LfU	LANDESANSTALT FÜR UMWELTSCHUTZ (KARLSRUHE,D)	INSTI
LGF		LARGE GENE FRAGMENTS	CLINCHEM
LH		LUTEINIZING HORMONE	CLINCHEM
LH-RH		LULIBERIUM-RELEASING HORMONE (=GONADOTROPES-RH)	CLINCHEM
LHASA		LOGIC AND HEURISTICS APPLIED TO SYNTHETIC ANALYSIS	METH
LHC		LIGHT HARVESTING COMPLEXES	COMP
LHCM		LIGHT HARVESTING COMPLEX MEMBRAN BOUNDED	COMP

Abkürzung Akromym	Alternative	Bedeutung	Zuordnung
abbreviation akronym	alternative	meaning	related section
LHCP		LIGHT HARVESTING CHLOROPHYLL ALBPROTEIN	MATER
LHKW		LEICHTFLÜCHTIGE HALOGENKOHLENWASSERSTOFFE	COMP
LIA		LUMINESCENCE IMMUNO ASSAY	METH
LIA		LOCK-IN AMPLIFIER	INSTR
LIB		LANDESANSTALT FÜR IMMISSIONS- UND BODENNUTZUNGSSCHUTZ (NRW,D)	INSTI
LIBS		LASER INDUCED BREAKDOWN SPECTROSCOPY	METH
LIDAR	Lidar	LIGHT DETECTION AND RANGING	METH
LIDUM		LITERATURINFORMATIONSDIENST UMWELT (UBA,D)	LIT
LIF		LASER INDUCED FLUORESCENCE	METH
LIFLN		LASER INDUCED FLUORESCENCE LINE NARROWING	INSTR
LIMA		LASER IONIZATION MASS ANALYSER/ SPECTROMETRIC ANALYSIS	INSTR METH
LIMFS		LASER INDUCED MOLECULAR FLUORESCENCE SPECTROMETRY	METH
LIMS		LABORATORY INFORMATION MANGEMENT SYSTEM	INSTR ELECT
LIS		LASER INDUCED ISOTOPE SEPARATION	METH
LIS		LANDESANSTALT FÜR IMMISSIONSSCHUTZ ESSEN (NRW,D)	INSTI
LLC		LIQUID LIQUID CHROMATOGRAPHY	METH
LLD		LOWER LIMIT OF DETECTION	EVAL
LLD		LACTOBACILLUS LACTIS DORNER FACTOR	CLINCHEM
LLDPE		LOW PRESSURE LOW DENSITY POLYETHYLENE (=LINEAR LOW DENSITY PE)	MATER POLYM
LLE	LLP	LIQUID LIQUID EXTRACTION	METH
LLP	LLPC	LIQUID LIQUID PARTITION & CHROMATOGRAPHY	METH
LLW		LOW LEVEL ACTIVE / RADIOACTIVE WASTE	TECHN
LM		LIGHT MICROSCOPY	METH
LM-TEM		LIGHT AND TRANSMISSION ELECTRON MICROSCOPE	INSTR
LMBG		LEBENSMITTEL UND BEDARFSGEGENSTAENDE GESETZ (D)	REG
LMC		LIGTH WIGHT COATING	TECHN
LMF		LEAF MOVEMENT FACTORS	BIOS

Abkürzung Akromym	Alternative	Bedeutung	Zuordnung
abbreviation akronym	alternative	meaning	related section
LMHV		LEBENSMITTEL HYGIENE VERORDNUNG (D)	REG
LMI		LIQUID METAL FIELD IONIZATION	INSTR
LMIS		LIQUID METAL ION SOURCE	INSTR
LMKV		LEBENSMITTEL KENNZEICHNUNGS VERORDNUNG (D)	REG
LMO		LOCALIZED MOLECULAR ORBITAL	THEOR
LMR		LASER MAGNETIC RESONANCETECHNIQUE	METH
LMSA		LE GRESSUS MASSIGNOTSOPIRET AUGER	THEOR
LMW		LOW MOLECULAR WEIGHT	COMP POLYM
LMWP		LOW MOLECULAR WEIGHT PROTEASE	CLINCHEM
LN2		LIQUID NITROGENE	MATER
LNA		LABORATORIUM DE NEDERLANDSE APOTHEKERS (NDL)	INSTI
LNG		LIQUID NATURAL GAS	MATER TECHN
LNU		LANDESGEMEINSCHAFT FÜR NATURSCHUTZ UND UMWELT e.V.	ORGANIS
LOD	NWG	LIMIT OF DETECTION	METH
LOES		LASER OPTICAL EMISSION SPECTROCOPY	METH
LOI		LOSS ON IGNITION	INSTR
LOQ		LIMIT OF QUANTIFICATION	METH
LPC		LOW PERFORMANCE CHROMATOGRAPHY	METH
LPC		LIQUID PHASE CATALYSIS	EXPER TECHN
LPCE		LIQUID PHASE CATALYTIC EXCHANGE	TECHN
LPF		LOW PASS FILTER	INSTR
LPG		LIQUIFIED PETROLEUM GAS	MATER TECHN
LPH		LIPOTROPIC HORMONE (ALPHA,BETA & GAMMA)	CLINCHEM
LPM		LASER MICROPROBE	INSTR
LPO		LAUROYL PEROXIDE	COMP
LPS		LIPO POLYSACCHARIDE	MATER
LRMA		LASER RAMAN MICRO ANALYSES	METH
LRMS		LOW RESOLUTION MASS SPECTROMETRY	METH
LRO		LONG RANGE ORDER	THEOR
LRV		LUFTREINHALTEVERORDNUNG (CH)	REG
LS		LIQUID SCINTILLATION	INSTR
LS		LIGHT SCATTERING	INSTR
LS-TTL		LOW-POWER SCHOTTKY TRANSITOR LOGIC	ELECT
LSB		LAST SIGNIFICANT BIT	ELCT
LSC		LIQUID SOLID CHROMATOGRAPHY	METH

Abkürzung Akromym	Alternative	Bedeutung	Zuordnung
abbreviation akronym	alternative	meaning	related section
LSD		LYSERGSAEURE - DIETHYLAMID	COMP CLINCHEM
LSF		LIGHT STRUCK FLAVOR (BEER)	TECHN
LSF		LINE SPREAD FUNCTION	EVAL
LSI		LARGE SCALE INTEGRATION	EVAL INSTR
LSM		LASER SCANNING MICROSCOPE	INSTR
LSR		LANTHANIDE SHIFT REAGENT (NMR)	COMP
LT		LEUKOTRIEN	COMP
LTA		LEAD TETRAACETATE	COMP
LTA	LTB	LEUKOTRIEN A B	COMP
LTA		LOW TEMPERATURE ASHING	METH
LTC	CTD	11-TRANS OR CIS LEUKOTRIEN C & D	COMP
LTE		LOCAL THERMODYNAMIC EQUILIBRIUM	THEOR
LTEM		LOW TEMPERATURE FLUORESCENCE PHOSPHORESCENCE ELECTRON MICROSCOPY	METH
LTFS		LOW TEMPERARURE FLUORESCENCE	METH
LTH		LUTEOTROPIC HORMONE	CLINCHEM
LTM		LOW THERMAL MASS	MATER
LTMAC		DODECYLTRIMETHYLAMMONIUM CHLORIDE	COMP
LTR		LONG TERMINAL REPEAT	ELECT
LTWS	LTwS	LAGERUNG UND TRANSPORT WASSERGEFAEHRDENDER STOFFE (D)	REG TECHN
LTX		LAPHOTOXIN	CLINCHEM
LUA		LEBENSMITTEL UNTERSUCHUNGSAMT (D)	INSTI
LUFA		LANDWIRTSCHAFTS UNTERSUCHUNGS UND FORSCHUNGSANSTALT (D)	INSTI
LUMO		LOWEST UNOCCUPIED MOLECULAR ORBITAL	THEOR
LUO		LABORATORY UNIT OPERATION	LABOR
LUV		LARGE UNILAMMELLAR VESICLE	CLINCHEM
LWA		LANDESAMT FÜR WASSER UND ABFALL (NRW,D)	INSTI
LWC		LIGHT WEIGHT COATING	TECHN
LWL	LCF	LICHTWELLENLEITER	INSTR
LWR		LEICHTWASSER REAKTOR	TECHN
LWR		LASER WRITE READ	ELECT INSTR
LYS	Lys	LYSINE	COMP
LÖLF		LANDESANSTALT F.ÖKOLOGIE, LAND-& FORSTWIRTSCHAFT (RECKLINGHHAUSEN,D)	INSTI

Abkürzung Akromym	Alternative	Bedeutung	Zuordnung
abbreviation akronym	alternative	meaning	related section
MA		MUCONDIALDEHYDE	COMP
MAA		MUCON DIALDEHYDETETRAETHYL ACETAL	COMP
MAA		MENTHOXYACETIC ACID	COMP
MAA		METHYLACETO ACETATE	COMP
MAB		4-METHOXYAZOBENZENE	COMP
MAB	MAb(s)	MONOCLONAL ANTIBODY(IES)	CLINCHEM
MAB		MAN AND BIOSPEHRE	ENVIR
MAD		17a-METHYL-5-ANDROSTENE-3b,17b-DIOL	COMP
MAD		METHYL ALUMINIUM-BIS(2,6-DITERTBUTYL-4-METHYL PHENOXIDE)	COMP
MADU		2'-DEOXY-5-(BUTYLAMINO) URIDINE	COMP
MAED		MICRO-AREA ELECTRON DIFFRACTION	METH
MAES		MULTI ELEMENT ATOMIC EMISSION SPECTROMETRY	METH
MAFF		MINISTRY OF AGRICULTURE, FISCHERIES AND FOOD (US)	INSTI
MAG		MOBILE ABFALLGRUPPE BADEN-WÜRTTEMBERG (D)	INSTI
MAIA		MAGNETIC ANTIBODY IMMUNO ASSAY	METH
MAK		MAXIMALE ARBEITSPLATZ KONZENTRATION	REG
MAK	MAk-TECHN	MONOKLONALE ANTIKÖRPER	CLINCHEM
MAL	MAL-	METHOXYCARBONYLMALEIMIDE LIGAND	COMP
MAL		MOLECULAR ABSORPTION SPECTROMETRY WITH LINE SOURCE	METH
MAM	MAMACETAT	(METHYLAZOXY)-METHANOL	COMP
MAM	6-MAM	6-MONOMACETYL MORPHINE	COMP CLINCHEM
MANEB	Maneb	ETHYLENE BIS(DITHIOCARBAMATO) MANGANESE	COMP
MAO		MONOAMINOOXIDASE	CLINCHEM
MAOT		MAXIMUM ALLOWABLE OPERATING TEMPERATURE	TECHN
MAP		MESSENGER ACTIVATED PAPER	MATER METH
MAP		MANUFACTURING AUTOMATION PROTOCOL	TECHN
MAP	MAP's	MICRO TUBULE-ASSOCIATED PROTEIN(S)	CLINCHEM
MAPA		N-METHYL-1,3-PROPANEDIAMINE	COMP
MAPO		TRIS-(1-(2-METHYL)AZIRINIDYL)PHOSPHINE OXIDE	COMP
MAPS		TRIS-(1-(2-METHYL)AZIRIDINYL)PHOSPHINE SULFIDE	COMP

Abkürzung Akromym abbreviation akronym	Alternative alternative	Bedeutung meaning	Zuordnung related section
MAPTAC		METHACRYLAMIDO PROPYL-TRIMETHYL AMMONIUM CHLORIDE	COMP
MAS		MAGIC-ANGLE SPINNING (NMR)	METH
MAS		MOLECULAR ABSORPTION SPECTROMETRY	METH
MAS		9,10-DIMETHOXY ANTHRACENE-2-SULFONATE	COMP
MASC		METHYLALUMINIUM SESQUICHLORIDE	COMP
MASER		MICRO-WAVE AMPLIFICATION BY STIMULATED EMISSION OF RADIATION	INSTR
MATEC		MAXIMUM ACCEPTABLE TOXICANT CONCENTRATION	CLINCHEM
MATR		MULTIPLE ATTENUATED TOTAL REFLECTANCE (INFRARED)	METH
MAW		MIDDLE ACTIVE WASTE	TECHN
MBA		N,N'-METHYLENE-BIS-ACRYLAMIDE	COMP
MBA		4,4'METHYLENE BIS ANILINE	COMP
MBBA		N-(p-METHOXYBENZYLIDENE)-p-BUTYL-ANILINE	COMP
MBC		METHYL-2-BENZIMIDAZOLE CARBAMATE	COMP
MBE		MOLECULAR-BEAM EPITAXY	INSTR
MBK		MINIMALE BAKTERIZIDE KONZENTRATION	CLINCHEM
MBOCA		METHYLENE-BIS-(o-CHLOROR-ANILINE)	COMP
MBP		MYELUM BASIC PROTEIN	CLINCHEM
MBRS		MOLECULAR-BEAM REACTIVE SCATTERING	METH
MBS		m-MALEIMIDOBENZYL-N-HYDROXISUCCINIMID	COMP
MBS		METHYL-METHACRYLATE-BUTADIENE STYRENE	MATER POLYM
MBSS		MOLECULAR-BEAM SURFACE SCATTERING	METH
MBT		METHYLENE-BIS-THIOCYANATE	COMP
MBT		MERCAPTOBENZTHIAZOLE	COMP
MBTFA		N-METHYL-BIS-(TRIFLUOROACETEAMIDE)	COMP REAG
MBTH		3-METHYL-2-BENZOTHIAZOLON HYDRAZON (HYDROCHLORID)	COMP REAG
MBTS		MOVING BELT TRANSFER SYSTEM (MS)	INSTR
MBTS		MERCAPTOBENZTHIAZYL ETHER =2,2'-DITHIOBIS(BENZOTHIAZOLE)	COMP
MBZ	MBz ,MOB	4-METHOXYBENZYL (AND ...ALCOHOL)	COMP
MC		MULTICHROMATIC	INSTR
MC-SCF		MULTICONFIGURAL SELF CONSISTING FIELD	THEOR

Abkürzung Akromym	Alternative	Bedeutung	Zuordnung
abbreviation akronym	alternative	meaning	related section
MCA		MULTICHANNEL ANALYZER	INSTR
MCA	MAb	MULTI COMPONENT ANALYSIS	METH
MCA		MONOCLONAL ANTIBODY	CLINCHEM
MCA		MONOCHLOROACETIC ACID	COMP
MCA		3-METHYL CHOLANTHRENE	COMP
MCA		MANUFACTURING CHEMISTS ASSOCIATION (US)	ORGANIS
MCAC		METAL CHELATE AFFINITY CHROMATOGRAPHY	METH
MCD		MAGNETIC CIRCULAR DICHROISM	INSTR
MCD		MICRO COULOMETRIC DETECTOR	INSTR
MCDD		MONOCHLORO-p-DIBENZODIOXIN	COMP
MCDF		MONOCHLORO DIBENZOFURANE	COMP
MCE		MERCAPTO ETHANOL	COMP
MCFM		MICRON CUBIC FEET PER HOUR	EVAL
MCHO		MONOMERIC CARBOHYDRATES	COMP
MCL		MAGNETIC CORRECTED LAMP	INSTR
MCM		MARKOV CHAIN METHOD	THEOR
MCP		m-CRESOL PURPLE (INDI)	COMP
MCP		MULTI CHANNNEL PLATE DETECTOR	INSTR
MCPA		4-CHLORO-2-METHYL-PHENOXYACETIC ACID	COMP
MCPB		4-(2-METHYL-4-CHLOROPHENOXY) BUTYRICACID	COMP
MCPBA		m-CHLOROPERBENZOIC ACID	COMP
MCPCA		2-METHYL-4-CHLOROPHENOXYACETO-o-CHLOROANILIDE	COMP
MCPDEA		N,N-DI-(2-HYDROXYETHYL)-m-CHLORO-ANILINE	COMP
MCPP		4-CHLORO-3-METHYLPHENOXYPROPIONIC ACID	COMP
MCS		MOISTURE CONTROL SYSTEM	INSTR
MCS		MULTI CHANNEL SPECTROMETER	INSTR
MCSCF	MC-SCF	MULTICONFIGURATION SELF CONSISTING FIELD	THEOR
MCT		MERCURY CADMIUM TELLURIDE DECTOR (INFRA RED)	INSTR
MD		MOLECULAR DYNAMICS	THEOR
MDA		METHYLENEDIOXY AMPHETAMINE	COMP CLINCHEM
MDA		MAGNETIC DEFLECTION ANALYZER	INSTR

Abkürzung Akromym	Alternative	Bedeutung	Zuordnung
abbreviation akronym	alternative	meaning	related section
MDA		1,8-p-MENTHANE DIAMINE	COMP
MDC	NWG	MINIMUM DETECTABLE CONCENTRATION	EVAL
MDEA		METHYL DIETHANOL AMIN	COMP
MDEB	-MDEB	(-)N-DODECYL-N-METHYL-EPHEDRINIUM BROMID	COMP
MDF		MICRO DOSE FOCUSING	INSTR
MDGC		MULTI DIMENSIONAL GASCHROMATOGRAPHY	METH
MDH		MALATE DEHYDROGENASE	CLINCHEM
MDH		METHANOL DEHYDROGENASE	CLINCHEM
MDHS		METHODS FOR THE DETERMINATION OF HAZARDOUS SUBSTANCES (GB)	NORM
MDI	4,4'MDI	METHYLENE DIPHENYL ISOCYANAE	COMP
MDIU	4,4'MDIU	4,4'DIPHENYL-DI(3-n-PROPYL-3(4-NITROBENZYL)UREA	COMP
MDL	LOD, NWG	MINIMUM DETECTABLE/ DETECTION LIMIT	EVAL METH
MDM		MINIMUM DETECTABLE MASS	EVAL
MDPF		2-METHOXY -2,4-DIPHENYL-3(2H)-FURANONE	COMP
MDQ		MINIMUM DETECTABLE/ DETECTION QUANTITY	EVAL
MDRI		MULTI DETECTOR RETENTION INDEX	EVAL
MDSN		MAXIMUM DISSOLVED SOLIDS NEBULIZER	INSTR
MEA		MOISTURE EVOLUTION ANALYSIS	METH
MEA		BETA-MERCAPTO ETHYLAMINE	COMP
MEC		MINIMUM ENERGY CONFORMER	THEOR
MECA		MOLECULAR EMISSION CAVITY ANALYSIS	METH
MED		MICROWAVE PLASMA EMISSION DECTOR	INSTR
MEDGV	MedGV	VERORDNUNG ÜBER DIE SICHERHEIT MEDIZINISCH TECHNISCHER GERÄTE (D)	REG
MEDLARS		MEDICAL LITERATURE ANALYSIS AND RETRIEVAL SYSTEM (INT)	LIT
MEEC		MEMBRANE-ENCLOSED ENZYMATIC CATALYSIS	TECHN
MEED		MEDIUM ENERGY ELECTRON DIFFRACTION	METH
MEHQ		METHYLHYDROQUINONE	COMP
MEI		2-MORPHOLINOETHYL ISOCYANIDE	COMP
MEIS		MEDIUM ENERGY ION SCATTERING SPECTROSCOPY	METH
MEK		METHYL-ETHYLKETON	COMP

Abkürzung Akromym / abbreviation akronym	Alternative / alternative	Bedeutung / meaning	Zuordnung / related section
MELF		MINISTERIUM FÜR ERNÄHRUNG, LANDWIRTSCHAFT UND FORSTEN (NRW,D)	INSTI
MEM	MEM-	METHOXYETHOXY METHYL LIGAND	COMP
MEM		MIRROR ELECTRON SPECTROSCOPY	METH
MEMO		3-METHACRYLOXYOXYPROPYL-TRIMETHOXY SILANE	COMP
MEO-PMS	1-MEO-PMS	1-METHOXY-5-METHYLPHENAZINIUM METHYL SULFATE	COMP
MEP		O,O-DIMETHYL O-(3-METHYL-4-NITROPHENYL)-PHOSPHOROTHIONATE	COMP
MERS		MULTIPLE ELECTRON RESONANCE SPECTROSCOPY	METH
MES		2-MORPHOLINO ETHANSULFONSAEURE	COMP
MES	MES-	METHANE SULFANYL-LIGAND	COMP
MES	MOSS	MOESSBAUER EFFECT SPECTROCOPY	METH
MES		MOLECULE EMISSION SPECTROMETRY	METH
MES-CL	MES-Cl	METHANESULFANYLCHLORIDE	COMP
MESAEP		MEDITERRANEAN SCIENTIFIC ASSOCIATION OF ENVIRONMENTAL PROTECTION (INT)	ORGANIS
MET	Met	METHIONINE	COMP
METS		METHANEPHRINES	COMP
MF		N-METHYLFORMAMID	COMP
MFD		MULTIPLE FUNCTION DETECTOR	INSTR
MFE		MERCURY FILM ELECTRODE	INSTR
MFHA		N-METHYLFUOROHYDROXAM ACID	COMP
MFI		MELTING FLOW INDEX	TECHN
MFLOPS		MILLION FLOATING OPERATIONS PER SECOND (NMR)	EVAL
MFO		(MICROSOMAL) MIXED FUNCTION OXIDASE	CLINCHEM
MFS		MOLECULE FLUORESCENCE SPECTROMETRY	METH
MGDG		MONOGALACTOSYLDIGLYCERIDE	COMP
MGF		MEDIUM SIZED GEN FRAGMENT(S)	CLINCHEM
MGH		MASSACHUSETTS GENERAL HOSPITAL (US)	INSTI
MHB		METHYL HYDROXY BENZOATE	COMP
MHC		MAJOR HISTOCOMPATIBILITY COMPLEX	CLINCHEM
MHD		MAGNETO HYDRODYNANICS	INSTR
MHH		MEDIZINISCHE HOCHSCHULE HANNOVER (D)	INSTI
MHHPA		METHYLHEXAHYDROPHTHALIC ANHYDRIDE	COMP
MHK		MINIMALE HEMMER KONZENTRATION	CLINCHEM

Abkürzung Akromym abbreviation akronym	Alternative alternative	Bedeutung meaning	Zuordnung related section
MHPG		3-METHOXY-4-HYDROXYPHENYLGLYCOL	COMP
MHW		MINISTRY OF HEALTH AND WELFARE (JAP)	INSTI
MIA		N-METHYLISATOIC ANHYDRIDE	COMP
MIBK	MIK	METHYL ISOBUTYL KETON	COMP
MIC	MHK	MINIMAL INHIBITORY CONCENTRATION	CLINCHEM
MID		MULTIPLE ION DETECTION	INSTR
MIF	MRIH	MELANOSTATIN	CLINCHEM
MIH	MSH	MELANOSTATIN (=MELANO TROPIN INHIBITING HORMONE)	CLINCHEM
MIK		MAXIMALE IMMISSIONS KONZENTRATION (D)	REG
MIKE		MASS ANALYZED ION KINETIC ENERGY	INSTR
MIKES		MASS ANALYZED ION KINETIC ENERGY SPECTROMETRY	METH
MIMO		MULTIPLE INPUT-MULTIPLE OUTPUT	TECHN
MINDAP	MINDAP-AES	MICROWAVE INDUCED NITROGEN DISCHARGE	INSTR
MINDO		MODIFIED INTERMEDIATE NEGLECT OF DIFFERENTIAL OVERLAP	THEOR
MINERVE		MANGEMENT INTERNATIONAL OF EUROPEAN RESIDUES AND WASTES	PROJ
MINTECH		DATENBANK EMISSIONSMINDERUNGSTECHNIK	LIT
MIP		MICROWAVE INDUCED PLASMA	INSTR
MIPK		METHYL ISOPROPYL KETONE	COMP
MIR		MULTIPLE INTERNAL REFLECTION	INSTR
MIR		MAXIMALE IMMISSIONSRATE	TECHN
MIR		MEDIUM RANGE INFRARED SPECTROSCOPY (400-200/CM)	METH
MIS		MULLERIAN INHIBITING SUBSTANCE(S)	CLINCHEM
MIT		MASSACHUSETTS INSTITUTE OF TECHNOLOGY BOSTON (US	INSTI
MITC		METHYLISOTHIOCYANATE	COMP
MITI		MINISTRY OF INTERNATIONAL TRADE AND INDUSTRY (J)	INSTI
MIX		3-ISOBUTYL-1-METHYLXANTHINE	COMP
MLA		MOELLENSTEDT LENS ANALYZER	INSTR
MLC		MOLECULAR LUMINESCENCE SPECTROMETRY WITH CONTINUUM SOURCE	METH
MLD		MINIMALE LETALE DOSIS =MINIMUM LETHAL DOSE	CLINCHEM

Abkürzung Akromym / abbreviation akronym	Alternative / alternative	Bedeutung / meaning	Zuordnung / related section
MLL		MOLECULAR LUMINESCENCE SPECTROMETRY WITH LINE SOURCE	METH
MLM		MONOLAYER LIPID MEMBRANE	MATER CLINCHEM
MLR		MULTIPLE LINEAR REGRESSION	EVAL
MLUA		MEDIZINISCHES LANDESUNTERSUCHUNGSAMT (D)	INSTI
MLV		(SMALL)MULTI UNILAMELLAR VESICLE	CLINCHEM
MLV		MOLOREY LEUKAEMIE VIRUS	CLINCHEM
MLW		MEDIUM LEVEL RADIOACTIVE WASTE	TECHN
MMA		METHYLMALONIC ACID	COMP
MMA		MONOMETHYL ARSONIC ACID	COMP
MMAA		MONO-N-METHYLACETOACETAMIDE	COMP
MMC		MITO MYCIN C	CLINCHEM
MMH		METHYLMERCURIC HYDROXIDE	COMP
MMM		MODEL MICROFIELDMETHOD (SPECT)	METH
MMPD	2,4-DAA	4-METHOXY-1,3-BENZENEDIAMINE	COMP
MMPO		MONOMETHYL PHOSPHATE	COMP
MMS		METHYL METHAN SULFONAT	COMP
MMS		MICROMEMBRANE SUPPRESSOR	INSTR
MMS		MULTIPHOTON MASS SPECTROMETRY	METH
MMT		MONOMETHYL TEREPHTHALATE	COMP
MMT		ALPHA-METHYL-m-TYROSINE	COMP
MMTS		METHYL METHYLSULFINYL METHYL SULFIDE	COMP REAG
MMTV		MOUSE-MAMMARY TUMOR VIRUS	CLINCHEM
MN		MACHEREY-NAGEL & CO	COMPANY
MNA		METHYLNORBORNEN-2,3-DICARBONSAEURE ANHYDID	COMP
MNDO		MODIFIED NEGLECT OF DIATOMIC OVERLAP	THEOR
MNNG		N-METHYL-N-NITROSO GUANIDINE	COMP
MNP		m-NITROPHENOL =3-NITROPHENOL	COMP
MNPA		6-METHOXY-a-METHYL-2-NAPHTHALENE ACETIC ACID	COMP
MNPT		m-NITRO-p-TOLUIDINE	COMP
MNSL		MAX. NITRIFIKATIONS SAUERSTOFFVERBRAUCH UNTER STANDARDLABORBEDING	METH TECHN
MNU		METHYL NITROSO UREA	COMP

Abkürzung Akromym abbreviation akronym	Alternative alternative	Bedeutung meaning	Zuordnung related section
MO		MOLECULAR ORBITAL	THEOR
MO-CVD		METALORGANICS CHEMICAL VAPOR DEPOSITION	TECHN
MOB	MOB-	4-METHOXYBENZYL-LIGAND	COMP
MOC		LIGAND FROM METHYL CHLOROFORMATE	COMP
MOCA	DACPM	4,4-METHYLENE-BIS(o-CHLOROANILINE) =4,4-METHYLENE-BIS(2-CHLOROANILINE	COMP
MOG		MILLS, OLNEY AND GAITHER PROCEDURE	THEOR
MOL		MINISTRY OF LABOUR (JAP)	INSTI
MOLE		MOLECULAR OPTICS LASER EXAMINER =RAMAN MICROPROBE	INSTR
MOM	MOM-	METHOXYMETHYL-LIGAND	COMP
MOMO		MAXIMUM OVERLAP MOLECULAR ORBITAL	THEOR
MOMS		MODULAR OPTOELECTRONICAL MULTIWAVELENTH SCANNER	INSTR
MOP	8-MOP	METHOXYPSORALEN	COMP
MOPAP		4-(1,1,3,3-TETRABUTYL) PHENYLDIHYDROGENPHOPHATE	COMP
MOPS		2-(N-MORPHOLINO)PROPANE SULFONIC ACID	COMP
MOPSO	MPS	3-(N-MORPHOLINO)-2-HYDROXYPROPANESULFONIC ACID	COMP
MOR		MAGNETO-OPTICAL ROTATION SPECTROSCOPY	METH
MORD		MAGNETIC-OPTICAL-ROTATIONS-DISPERSION	METH
MOS		METAL OXIDE SEMICONDUCTOR	ELECT MATER
MOSFET		METAL OXIDE SEMICONDUCTOR FIELD EFFECT TRANSISTOR	ELCHEM
MOSS	MES	MOESSBAUER SPECTROSCOPY	METH
MOZ	MOZ-	4-METHOXYBENZYL-LIGAND	COMP
MOZ	Moz	MOTOR OCTANZAHL	TECHN
MP	MP2,3,4	MOLLER PLESSET 2 ND ORDER	THEOR
MPA	12-MPA	12-MOLYBDATO PHOSPHATE	COMP
MPA		MULTIPHOTON ABSORPTION	METH
MPA		3-MERCAPTO PROPIONIC ACID	COMP
MPA		MATERIALPRÜFUNGSANSTALT STUTTGART (D)	INSTI
MPB		2-MERCAPTO-1(2-(4-PYRIDYLETHYL)-BENZIMIDAZOLE	COMP

Abkürzung Akromym / abbreviation akronym	Alternative / alternative	Bedeutung / meaning	Zuordnung / related section
MPCA	MPCa	MAXIMUM PERMISSILE CONCENTRATION IN AIR	REG
MPCW	MPCw	MAXIMUM PERMISSIBLE CONCENTRATION IN WATER	REG
MPD		MULTI-PHOTON DISSOCIATION	INSTR
MPDC		N-METHYL-3-PIPERIDYL-(N',N')-DIPHENYLCARBAMATE	COMP
MPE		MATHEMATICAL AND PHYSICAL SCIENCE AND ENGENEERING (NSF/US)	INSTI
MPE		MULTIPHOTON EXCITATION	INSTR
MPEMA		2-ETHYL-2-(p-TOLYL) MALONAMIDE	COMP
MPG		MAX PLANCK GESELLSCHAFT ZUR FOERDERUNG DER WISSENSCHAFT (D)	INSTI
MPG		METAPHOSPHATGLAS	MATER TECHN
MPI		MULTIPHOTON IONIZATION SPECTROSCOPY	METH
MPI		MAX-PLANK-INSTITUT (MPG,D)	INSTI
MPIC		MOBILE PHASE ION CHROMATOGRAPHY	METH
MPLC		MEDIUM PRESSURE LIQUID CHROMATOGRAPY	METH
MPM		MULTIPLE PEAK MONITORING	INSTR
MPML		METHOYX (PHENYLTHIO) METHYL LITHIUM	COMP
MPN		2-METHOXYPHENYL-AZO-1-(2-HYDROXYNAPHTHYL-7-TRIMETHYLAMMONIUM)CHLORIDE	COMP
MPOSS		MULTI PURPOSE OILSKIMMER SYSTEM	ENVIR TECHN
MPP		O,O-DIMETHYL O-(4-METHYLMERCAPTO-3-METHYLPHENYL) THIOPHOSPHATE	COMP
MPPH		5-METHYLPHENYL-5-PHENYL HYDANTOIN	COMP
MPS	MOPS	3-MORPHOLINO PROPAN SULFONSAEURE	COMP
MPS		METHYL PHENYL SULFIDE	COMP
MQ		METHYLSILICONRUBBER	MATER
MQ		MEDIUM QUANTITY (MITTLERE FRACHT)	TECHN
MQSG		2-METHYL-8-QUINOLINOL SILICA GEL	MATER
MRA		MIXED REFRIGERANT AUTO-CASCADE	TECHN
MRC		MEDICAL RESEARCH COUNCIL (GB)	ORGANIS
MRC		MORAL RISK CALCULATION	EVAL
MRH		MELANOLIBERIN (=MELANOTROPIN RELEASING HORMONE)	CLINCHEM
MRH	Mrh	MANNOSE RESISTENT HEMAGGLUTININE	CLINCHEM

Abkürzung Akromym abbreviation akronym	Alternative alternative	Bedeutung meaning	Zuordnung related section
MRI		MAGNETIC RESONANCE IMAGING	METH
MRI		MICROBIOLOGICAL RESEARCH INSTITUTE OTTAWA (CAN)	INSTI
MRIH	MIF	MELANOCYTE-STIMULATING HORMONE RELEASE INHIBITING FACTOR	CLINCHEM
MRITC		METHYL-RHODAMINE ISOTHIOCYANATE	COMP
MROA		MAGNETIC RAMAN OPTICAL ACTIVITY	INSTR
MRS	MRS-Agar	DE MAN,ROGASA SHARP AGAR	CLINCHEM
MRS		MSH-RELEASING HORMONE	CLINCHEM
MRS		MICRO RAMAN SPECTROSCOPY	METH
MS	MOSS	MOESSBAUER SPECTROSCOPY	METH
MS	MS-MODEL	MARTIN-SYNGE MODELL IN CHROMATOGRAPHY	THEOR
MS		MULTIPLE SCLEROSIS	CLINCHEM
MS		MASS SPECTROMETRY /MASS SPECTROSCOPY	METH
MS2	MSMS	MASS SPECTROMETRY COUPLED MASS SPECTROMETRY	METH
MSA	12-MSA	12-MOLYBDATOSILICATE	COMP
MSA		N-METHYL-N-TRIMETHYLSILYLACETAMIDE	COMP
MSA		METHANESULFONIC ACID	COMP
MSCL	MSCI	MESITYLENE-2-SULFONYL CHLORIDE	COMP
MSD		MEDIZINISCH PHARMAZEUTISCHE STUDIENGESELLSCHAFT (D)	INSTI
MSD	MSDRL	MERCK,SHARP AND DOHME /RESEARCH LABORATORY	COMPANY
MSD		MASS SPECTROMETRIC DETECTOR	INSTR
MSF	MSF/FBE	MULTISTAGE FLASH DISTILLATION/FLUID BED EXCHANGER	TECHN
MSH		o-MESITYLENESULFENYLHYDROXYLAMINE	COMP
MSH		MELANOPHORE EFFECTING HORMONE =MELANOSTATIN	CLINCHEM
MSH	Msh	MANNOSE SENSITIVE HEMAGGLUTININE	CLINCHEM
MSHA		MINING SAFETY HEALTH ACT (US)	REG
MSI		MEDIUM SCALE INTEGRATION	ELCT
MSID	IDA	MASS SPECTROMETRIC ISOTOPE DILUTION	METH
MSMA		MONO-SODIUM METHANARSONATE	COMP
MSNT		(MESITYLENE-2-SULFONYL)-3-NITRO 1,2,3-TRIAZOLE	COMP
MSO		p-CRESYL METHYL ETHER	COMP

Abkürzung Akromym / abbreviation akronym	Alternative / alternative	Bedeutung / meaning	Zuordnung / related section
MSOC		2-METHYLSULFONYL-ETHYL-4-NITROPHENYL CARBONAT	COMP REAG
MSR		MULTIPLE SPECTULAR REFLECTANCE	INSTR
MSRTP		MICELLE-STABILIZED ROOM TEMPERATURE PHOSPHORESCENCE	METH
MSSS		MASSSPECTROMETRY SEARCH SYSTEM	INSTR
MST		MEAN SURVIVAL TIME	CLINCHEM
MSTE	MSTe	1-MESITYLENE-SULFONYL-TETRAZOLE	COMP REAG
MSTFA		N-METHYL-N-TRIMETHYLSILYL-TRIFLUOROACETAMIDE	COMP REAG
MSV		MOLOREY SARKOM VIRUS	CLINCHEM
MT	5-MT 3-MT	METHOXY TYRAMINE DIV.	COMP
MT		DL-ALPHA-METHYLTYROSINE	COMP
MT-HF	5-MT-HF	5-METHYL TETRAHYDROFOLIC ACID	COMP
MTB	MTBE	METHYL TERT-BUTYLETHER	COMP
MTBSTFA		M-METHYL-N-TERT-BUTYLDIMETHYLSILYL TRIFLUOROACETAMIDE	COMP RAG
MTC	MITC	METHYL ISOTHIOCYANATE	COMP
MTCA		2-METHYLTHIAZOLIDINE-4-CARBOXYLIC ACID	COMP
MTCD		MICRO VOLUME THERMAL CONDUCTIVITY DETECTOR	INSTR
MTD		m-TOLUENE-DIAMINE	COMP
MTDEA		N,N-DI(2-HYDROXYETHYL)-m-TOLUIDINE =m-TOLUIDINE-N,N-DIETHANOL	COMP
MTES		METHYL-TRIETHOXY SILANE	COMP
MTFE		MERCURY THIN FILM ELECTRODE	ECHEM INSTR
MTG		METHYL BETA-D-THIOGALACTOSIDE	COMP
MTH		METHYLTHIOHYDANTOIN	COMP
MTHF		2-METHYL TETRAHYDROFURAN	COMP
MTHPA		METHYL-TETRAHYDROPHTHALIC ANHYDRIDE	COMP
MTM	MTM-	METHYL-THIOMETHYL-LIGAND	COMP
MTMC		4-(METHYLTHIO)-m-CRESOL	COMP
MTMS		METHYL-TRIMETHOXY-SILANE	COMP
MTN		m-TOLYLNITRILE	COMP
MTO	OMD	3-METHOXY-L-TYROSINE	COMP
MTP		METHYLTRIPHENOXYPHOSPHONIUM -SALT	COMP
MTP		4-(METHYLTHIO)-PHENOL	COMP
MTPA		ALPHA-METHOXY-ALPHA-TRIFLUOROMETHYL PHENYL ACETIC ACID	COMP

Abkürzung Akromym abbreviation akronym	Alternative alternative	Bedeutung meaning	Zuordnung related section
MTPI		METHYLTRIPHENOXY PHOSPHONIUM IODIDE	COMP
MTQ		METHAQUALANE	COMP
MTS		MOLECULAR TRANSMISSION SPECTROMETRY	METH
MTT		(3-(4,5-DIMETHYL THIAZOL-2YL)-2,5-DIPHENYL-2H-TETRAZOLIUM SALT	COMP REAG
MTT	MTT-COMPL	(MANGAN) METHYLCYCLOPENTADIENYL-TRICARBONYL	COMP
MTX		METHOTREXATE	COMP
MU	4-MU	4-METHYL UMBELLIFERONE	COMP
MUGB		4-METHYL-UMBELLIFERYL p-GUADINOBENZOATE	COMP
MVA		MÜLLVERBRENNUNGSANLAGE	TECHN
MVI		MELTING VOLUME INDEX	TECHN
MVK		METHYL VINYL KETONE	COMP
MVP		2-METHYL-5-VINYLPYRIDINE	COMP
MVT		MONOMETHYL VINYL TEREPHTHALATE	COMP
MW		MOLECULAR WEIGHT =MOLARE MASSE	COMP
MWD		MOLECULAR WEIGHT DISTRIBUTION	MATER
MWFT		MICROWAVE FOURIER TRANSFORM	SPECT
MWP		MICROWAVE PLASMA	INSTR
MWS		MICROWAVE SPECTROSCOPY	METH
MWV		MINERALÖLWIRTSCHAFTSVERBAND HAMBURG (D)	ORGANIS
MZV		MINERALÖL ZENTRALVERBAND (D)	ORGANIS

Abkürzung Akromym	Alternative	Bedeutung	Zuordnung
abbreviation akronym	alternative	meaning	related section
NAA		NEUTRON ACTIVATION ANALYSIS	METH
NAA		1-NAPHTHYLACETIC ACID	COMP
NAAA		NATIONAL AGRICULTURAL AVIATION ASSOCIATION (US)	ORGANIS
NAAD		NICOTINIC ACID ADENINE DINUCLEOTIDE	COMP CLINCHEM
NAC		N-ACETYL-L-CYSTEIN	COMP
NAC		1-NAPHTHYL N-METHYLCARBAMATE	COMP
NACA		NATIONAL AGRICULTURAL CHEMICALS ASSOCIATION (US)	ORGANIS
NACD		NATIONAL ASSOCIATION OF CHEMICAL DISTRIBUTORS (US)	ORGANIS
NACOSH		NATIONAL ADVISORY COMMITTEE ON OCCUPATIONAL SAFETY AND HEALTH (US)	INSTI
NAD	DPN	NICOTINAMIDE ADENOSINE DINUCLEOTIDE	COMP CLINCHEM
NADH		NICOTINAMIDE ADENEINE DINUCLEOTIDE PHOSPHATE REDUCED	COMP CLINCHEM
NADL		NATIONAL ANIMAL DISEASE LABORATORY AMES (US)	INSTI
NADP		NICOTINAMIDE ADENOSINE DINUCLEOTIDE PHOSPHATE	COMP
NAI		N-ACETYLIMIDAZOLE	COMP
NAM		N-ACETYLMETHIONINE	COMP
NAM		NATIONAL ASSOCIATION OF MANUFACTURERS (US)	ORGANIS
NAMUR		NORMEN-AG FÜR MEß-UND REGELUNGSTECHNIK IN DER CHEMISCHEN INDUSTRIE (D	ORGANIS
NANA		N-ACETYLNEURAMINIC ACID	COMP
NANB		NON A NON B HEPATITIS	CLINCHEM
NANM		N-ALLYL NORMORPHINE	COMP
NAP		4-NITRO AMINO PHENOL	COMP
NAPA	NAPAP	N-ACETYL-P-AMINOPHENOL	COMP
NAPA		N-ACETYL PROCAINAMIDE	COMP
NAPAP	NAPA	N-ACETYL PARA AMINOPHENOL	COMP
NAPCA	NAPCAC	NATIONAL AIR POLLUTION CONTROL ADMINISTRATION/ADVISORY COMMITTEE	INSTI
NAPE	N-APE	N-ACYLPHOSPHATIDYLETHANOL AMINE	COMP
NAPIM		NATIONAL ASSOCIATION OF PRINTING INK MANUFACTURERS	ORGANIS

Abkürzung Akromym abbreviation akronym	Alternative alternative	Bedeutung meaning	Zuordnung related section
NAPPB		NATIONAL COLLECTION OF PLANT PATHOGENIC BACTERIA HARPENDEN (GB)	INSTI
NAR		NORMENAUSSCHUSS RADIOLOGIE (DIN , D)	INSTI
NARP		NON AQUOS REVERSED PHASE	MATER
NAS		NATIONAL ACADEMY OF SCIENCE (US)	ORGANIS
NASA		NATIONAL AERONAUTICS AND SPACE ADMINISTRATION (US)	INSTI
NASCO		NORTH ATLANTIC SALMON CONSERVATION ORGANISATION (INT)	ORGANIS
NATEC		INSTITUT FUER NATURWISSENSCHAFTLICH-TECHNISCHE DIENSTE GmbH (D)	COMPANY
NATO-ASI		NORTH ATLANTIC TREATY ORGANIZATION-ADVANCED STUDY INSTITUTE	INSTI
NAW		NORMENAUSSCHUSS WASSERWESEN (DIN,D)	NORM
NAW		NON ACID WASHED	MATER
NAWDC		NATIONAL ASSOCIATION OF WASTE DISPOSAL CONTRACTORS (GB)	ORGANIS
NBA		N-BROM-ACETAMIDE	COMP
NBCA		4-(4-NONYLOXY BENZOYLOXY)-4'-CYANO AZOBENZENE	COMP
NBD		NORBORNADIEN	COMP
NBD		4-NITRO-2,1,3-BENZOXADIAZOLE	COMP
NBD-CL		7-CHLORO-4-NITROBENZO-2-OXA-1,3 DIAZOLE	COMP
NBD-F		4-FLUORO-NITROBENZO-2-OXA-1,3-DIAZOLE	COMP
NBDNO		3-BROMO-4,4-DIMETHYL-2-OXAZOLIDONE	COMP
NBE		TRANS-1-(4-BIPHENYL)-2-(1-NAPHTYL)ETHYLENE	COMP
NBP		4-(4-NITROBENZYL)-PYRIDINE	COMP
NBR		ACRYLNITRILE-BUTADIENE-RUBBER	MATER
NBS		NATIONAL BUREAU OF STANDARDS (US)	INSTI
NBS	NBS-	4-NITROBENZENESULFONATE-LIGAND	COMP
NBS		N-BROMO SUCCINIMDE	COMP
NBSAC	NBSac	N-BROMSACCHARIN	COMP
NBSC		2-NITROBENZENESULFENYL CHLORIDE	COMP
NBSG		NITROSOBENZAMIDE SILICAGEL	MATER
NBST	p-NBST	1-(4-NITROBENZOL(SULFANYL)-1H-1,2,4-TRIAZOL	COMP
NBT		NITROBLAU-TETRAZOLIUM CHLORID	COMP

Abkürzung Akromym abbreviation akronym	Alternative alternative	Bedeutung meaning	Zuordnung related section
NBTA		3-NITROBENZYL-TRIMETHYLAMMONIUM-LIGAND	COMP
NCA		N-CHLOROACETAMIDE	COMP
NCA		NATIONAL CANNER'S INSTITUTE WASHINGTON (US)	INSTI
NCCLS		NATIONAL COMMITTE FOR CLINICAL LABORATORY STANDARDS (US)	ORGANIS
NCDC		2-NITRO-4-CARBOXYPHENYL-N,N-DIPHENYLCARBAMATE	COMP
NCDO		NATIONAL COLLECTION OF DIARY ORGANISM SHINFIELD (GB)	INSTI
NCEC		NATIONAL CHEMICAL EMERGENCY CENTER (GB)	INSTI
NCI		NATIONAL CANCER INSTITUTE (US)	INSTI
NCI		NEGATIVE CHEMICAL IONIZATION	INSTR
NCIB		NATIONAL COLLECTION OF INDUSTRIAL BACTERIA ABBERDEEN (GB)	INSTI
NCIM		NATIONAL COLLECTION OF INDUSTRIAL MICROORGANISM POONA (IND)	INSTI
NCMB		NATIONAL COLLECTION OF MARINE BACTERIA ABERDEEN (GB)	INSTI
NCN		CYANONAPHTHALENE	COMP
NCRR		NATIONAL CENTER FOR RESOURCE RESEARCH (US)	INSTI
NCS		N-CHLORODICYANOBENZOQUINONE	COMP REAG
NCS		N-CHLOROSUCCINIMIDE	COMP
NCTC		NATIONAL COLLECTION OF TYPE CULTURES (LONDON,GB)	INSTI
NCTR		NATIONAL CENTER FOR TOXICOLOGICAL RESEARCH (US)	INSTI
NCYC		NATIONAL COLLECTION OF YEAST CULTURES NORWICH (GB)	INSTI
ND		N-NITROSODIETHANOLAMINE	COMP
ND		NEUTRON DIFFRACTION	INSTR
NDBA		N-NITROSO DIBUTYLAMINE	COMP
NDCEP	N-DCEP	N-(1,2-DICHLOROETHYL)PHTHALIMIDE	COMP
NDDO		NEGLECT OF DIATOMIC DIFFERENTIAL OVERLAP	THEOR
NDEA	DEN DENA	N-ETHYL-N-NITROSOETHAN AMINE	COMP
NDELA		N-NITROSODIETHANOL AMINE	COMP

Abkürzung Akromym	Alternative	Bedeutung	Zuordnung
abbreviation akronym	alternative	meaning	related section
NDGA		NORDIHYDROGUAIARETIC ACID	COMP
NDI		NEGLECT OF DIFFERENT (DIFFICULT) INTEGRALS	THEOR
NDIR		NONDISPERSIVE INFRARED	METH
NDMA		N-NITROSO DIMETHYLANILINE	COMP
NDO		NEGLECT OF DIFFERENTIAL OVERLAP	THEOR
NDOA		N-NITROSO-2(3',7'DIMETHYL-2',6'-OCTADIENYL)AMINOETHANOL	COMP
NDP		NEUTRON DEPTH PROFILING	METH
NDP		NEOPTERIN -3'-DIPHOSPHATE	COMP
NDP		1-(4-NITROBENZYL)-4- (4-DIETHYLAMINOPHENYLAZO) PYRIDINIUMBROMIDE	COMP
NDP		NUCLEOSIDE DIPHOSPHATE	COMP
NDPA		N-NITOSO DIPROPYLAMIN	COMP
NDS		NEUTRON DISPERSIVE SYSTEM	INSTR
NDT		NON DESTRUCTIVE TESTING	METH TECHN
NEA		NUCLEAR ENERGY AGENCY (OECD,INT)	INSTI
NEDA		N-1-NAPHTHYL ETHYLENE DIAMINE	COMP REAG
NEDSG		1-NAPHTHYL-ETHYLENE DIAMINE SILICAGEL	MATER
NEFF		NATIONALER ENERGIE FORSCHUNGS FOND (CH)	ORGANIS
NEFYTO		NEDERLAMDSE STICHTING VOOR FTYOFARMACIE	ORGANIS
NEI		NEGATIVE ION WITH ELECTRON IMPACT IONIZATION	EVAL
NEL	NOEL	NON EFFECT LEVEL	CLINCHEM
NEM		N-ETHYLMALEIMIDE	COMP
NEP		N-ETHYL-2-PYRROLIDONE	COMP
NEPA		NATIONAL ENVIORNMENT PROTECTION AGENCY (S)	INSTI
NEPIS		N-ETHYL-5-PHENYLISOXAZOLIUM-3'-SULFONATE	COMP
NET		NEGATIVE ENTROPY TRAP	INSTR
NET		NEXT EUROPEAN TORUS	TECHN
NEZ		NUKLEARES ENTSORGUNGS ZENTRUM (GEPLANT,D)	TECHN
NF		THE NATIONAL FORMULARY (US)	NORM
NF-AAS		NON FURNACE ATOMIC ABSORPTION	METH

Abkürzung Akromym	Alternative	Bedeutung	Zuordnung
abbreviation akronym	alternative	meaning	related section
NFPA		NATIONAL FIRE PREVENTION ASSOCIATION (US)	ORGANIS
NGA		NEOPENTYL GLYCEROL ADIPATE	MATER
NGF		NERVE GROWTH FACTOR	CLINCHEM
NGO		NON GOVERNMENTAL ORGANIZATION	ORGANIS
NGS		NEOPENTYL GLYCOL SEBATE	MATER
NHDC		NEOHESPERIDIN-DIHYDROCHALONE	COMP
NHDV		NORMALIZED HYDRODYNAMIC VOLTAMMOGRAMS	ELCHEM EVAL
NHE		NORMAL HYDROGENE ELECTRODE	ELCHEM INSTR
NHMRC		NATIONAL HEALTH AND MEDICAL RESEARCH COUNCIL	INSTI
NIA	5-NIA	5-NITROISATOIC ANHYDRIDE	COMP
NIA		NEDERLANDSE INDUSTRIE APOTHEKEER (KNMP,NL)	ORGANIS
NIAS		NATIONAL INSTITUTE OF AGRICULTURE SCIENCES NISHIGAHARA (JAP)	INSTI
NICHHD		NATIONAL INSTITUTE OF CHILD,HEALTH AND HUMAN DEVELOPMENT (US)	INSTI
NICI	NCI	NEGATIVE ION CHEMICAL IONIZTION	INSTI
NIDA		NATIONAL INSTITUTE FOR DRUG ABUSE (US)	INSTI
NIEHS		NATIONAL INSTITUTE OF ENVIRONMENTAL HEALTH SCIENCE (US)	INSTI
NIFAB		NORDISK INFORMATION FÖR FÄRG AB MALMÖ (S)	LIT
NIGMS		NATIONAL INSTITUTE OF GENERAL MEDICAL SCIENCE (US)	INSTI
NIH		NATIONAL INSTITUTE OF HEALTH (US)	INSTI
NIHS		NATIONAL INSTITUTE OF HYGIENE SCIENCE (JAP)	INSTI
NIIA		NON ISOTOPIC IMMUNO ASSAY	METH
NILU		NORWEGISCHES INSTITUT FÜR LUFTFORSCHUNG (N)	INSTI
NIM		NUCLEAR INSTRUMENT MODULES	INSTR TECHN
NIOSH		NATIONAL INSTITUTE OF OCCUPATIONAL SAFETY AND HEALTH (US)	INSTI
NIP		4-HYDROXY-5-NITRO-3-IODOPHENYLACETIC ACID	COMP
NIPHEGAL		2-NITROPHENYL-BETA-D-GALACTOPYRANOSID	COMP CLINCHEM

Abkürzung Akromym	Alternative	Bedeutung	Zuordnung
abbreviation akronym	alternative	meaning	related section
NIR	NIR-DR	NEAR INFRARED -DIFFUSE REFLECTION	INSTR
NIR		NEUTRAL IMPACT RADIATION	INSTR
NIRA		NEAR INFRARED REFLECTANCE ANALYSIS	METH
NIRMS		NOBLE GAS ION REFLECTION MASS SPECTROSCOPY	METH
NIRS		NEUTRAL IMPACT RADIATION SPECTROSCOPY	METH
NIRS		NEAR INFRARED SPECTROSCOPY	METH
NIS		N-IODO SUCCINIMIDE	COMP
NIS		NEGATIVE ION SPECTROSCOPY	METH
NKE	NKe	NORMENAUSSCHUSS KERNTECHNIK (DIN D)	NORM INSTI
NKI		NORGES KJEMZIKE INDUSTRIGRUPPE (OSLO,N)	ORGANIS
NLE	Nle	NORLEUCINE	COMP
NLLS		NONLINEAR LEAST SQUARES ANALYSIS	EVAL
NLM		NATIONAL LIBRARY OF MEDICINE (US)	LIT`
NMA		METHYL-NORBORNEN-2,3-DICARBONSAEUREANHYDRID	COMP
NMA		N-METHYLOL-ACRYLAMIDE	COMP
NMHC		NON METHAN HYDROCARBON	COMP
NMN	BETA-NMN	BETA-NICOTINAMID-MONONUCLEOTID	CLINCHEM
NMO		N-METHYLMORPHOLINE-N-OXIDE MONOHYDRATE	COMP
NMOR		N-NITROSO MORPHOLINE	COMP
NMP		NEOPTERIN-3'-MONOPHOSPHAT	COMP
NMP		N-METHYLPHTHALIMIDE	COMP
NMP		N-METHYLPYRROLIDONE (=1-METHYL-2-PYRROLIDONE)	COMP
NMP		NORMENAUSSCHUSS MATERIALPRÜFUNG (DIN - D)	NORM INSTI
NMP		NUCLEOSIDE MONOPHOSPHATE	COMP
NMR	KMR	NUCLEAR MAGNETIC RESONANCE	METH
NMSO		4-METHYL-NITRSOANISOLE	COMP
NNDO		NEGLECT OF NONBONDED DIFFERENTIAL OVERLAP	THEOR
NNI		NEDERLANDS NORMALISATIE-INSTITUT (NL)	INSTI
NNR		NEBENNIERENRINDE	CLINCHEM
NOAEL		NO OBSERVED ADVERSE EFFECT LEVEL	CLINCHEM

Abkürzung Akronym abbreviation akronym	Alternative alternative	Bedeutung meaning	Zuordnung related section
NODUS		NONDESTRUCTIVE AND ULTRASENSITIVE SINGLE ATOMIC LAYER SURFACE SPECTROSCOPY	METH
NOE		NUCLEAR OVERHAUSER ENHANCEMENT EFFECT (NMR)	EVAL
NOEC		NO OBSERVED EFFECT (LEVEL) CONCENTRATION	CLINCHEM
NOEL	NEL	NO OBSERVED EFFECT LEVEL	CLINCHEM
NOESY		NUCLEAR OVERHAUSER EFFECT SYSTEM	METH
NOP		NORMEN ORGANISATION PRAXIS (CH)	NORM
NOR		NORTRIPTYLINE	COMP
NOTA		1,4,7-TRIAZACYCLONONANE-N,N',N'' - TRIACETATE	COMP
NP		BIS (4-NITROPHENYL)DISULFIDE =NITROPHEMIDE	COMP
NP	1-NP	1-NITROPYRENE	COMP
NPA		p-NITRO PHENYLACETAT	COMP
NPA	NPA-ACID	N-(3-NITROFURFURYLLIDENE) CARBOXYL ARSANILIC ACID	COMP
NPCA		NATIONAL PAINT AND COATINGS ASSOCIATION (US)	ORGANIS
NPD		NITROGEN PHOSPHOROUS DETECTOR	INSTR
NPDES		NATIONAL POLLUTANT DISCHARGE ELIMINATION SYSTEM (US)	ENVIR
NPDL		NITROGEN PUMPED DYE LASER	INSTR
NPDPP	pNPDPP	P-NITROPHENYL DIPHENYL PHOSPHATE	COMP
NPG		NON PYROLYTIC GRAPHITE	MATER
NPGB		4-GUANIDINO-BENZOICACID-4-NITROPHENYLESTER HYDROCHLORIDE	COMP
NPIP		N-NITROSO PIPERIDINE	COMP
NPL		NATIONAL PHYSICAL LABORATORY (GB)	INSTI
NPLC		NORMAL PHASE LIQUID CHROMATOGRAPHY	METH
NPN		NON PROTEIN NITROGEN	CLINCHEM
NPO	ALPHA-NPO	2-(1-NAPHTHYL)-5-PHENYLOXAZOLE	COMP
NPOH		NONPURGEABLE ORGANIC HALOGENS	COMP
NPP		2-NITRO-2-PROPENYL-PIVALATE	COMP
NPS		o-NITROPHENYLSULFENYL-LIGAND	COMP
NPS	NPS-Cl	2-NITROPHENYL SULFENYL CHLORIDE	COMP
NPSD		NITROGEN-PHOSPHOR-SULFUR DETECTOR	INSTR

Abkürzung Akromym / abbreviation akronym	Alternative / alternative	Bedeutung / meaning	Zuordnung / related section
NPSP		N-PHENYLSELENENYLPHTHALIMIDE	COMP
NPT		NATIONAL PIPE TUPER (US)	INSTR NORM
NPYR		N-NITROSO PYRROLIDINE	COMP
NQHFS		NUCLEAR/NUCLEOUS QUADRUPOL HYPER FINE STRUCTURE	SPECT THEOR
NQR		NUCLEAR QUADRUPOLE RESONANCE	METH
NR		NATURAL RUBBER (=ISOPRENE RUBBER)	MATER
NRA		NUCLEAR REACTION ANALYSIS	METH
NRC		NATIONAL RESEARCH COUNCIL (US)	INSTI
NRC		NUCLEAR REGULATORY COMMISSION (US)	INSTI
NRCC		NATIONAL RESEARCH COUNCIL CANADA (CDN)	INSTI
NRDC		NATIONAL RESEARCH DEVELOPMENT CORPORATION (US)	INSTI
NRPB		NATIONAL RADIOLOGICAL PROTECTION BOARD (GB)	INSTI
NRS		NORMAL RAMAN SPECTROMETRY	METH
NSA		NICHT STEROIDALE ANTIPHLOGISTIKA	COMP CLINCHEM
NSAR		NICHT STEROIDALE ANTIRHEUMATIKA	COMP CLINCHEM
NSD		NITROGEN SELECTIVE DETECTOR	INSTR
NSER	n-Ser	n-SERINE	COMP
NSES		NEUTRON SPIN ECHO SPECTROSCOPY	METH
NSF		NATIONAL SCIENCE FOUNDATION (US)	INSTI
NSI		NATIONAL STANDARDS INSTITUTE (US)	INSTI
NSILA-S		NON SUPPRESIBLE INSULIN-LIKE ACTING SUBSTANCE	CLINCHEM
NSM		NEDERLANDSE STIKSTOFMIJ.NV (NL)	COMPANY
NTA		NITRILO TRIACETIC ACID	COMP
NTB	N-t-B	2-METHYL-2-NITOSOPROPANE (=NITROSO-tert-BUTANE)	COMP
NTC		NEGATIVE TEMPERATURE COEFFICIENT	EVAL
NTHC		NORGE TECHNICAL HOGSKOLES COLLECTION TRONDHEIM (N)	INSTI
NTI		NEGATIVE THERMIONIC MASS SPECTROMETRY	METH
NTIS		NATIONAL TECHNICAL INFORMATION SERVICE (US)	LIT
NTP		NEOPTERIN TRIPHOSPHAT	COMP
NTP		DESOXYRIBONULEOSID TRIPHOSPHATE DIVERSE	COMP CLINCHEM

Abkürzung Akromym abbreviation akronym	Alternative alternative	Bedeutung meaning	Zuordnung related section
NTU		NEPHELOMETRIC TURBIDITY TURBIDNESS UNITS	METH
NVA	Nva	NORVALINE	COMP
NVAOX		NON VOLATILE ADSORBABLE ORGANIC HALOGEN	COMP
NVFK		NEDERLANDSE VERENIGING-FEDERATIE VOOR KUNSTSTOFFEN (NL)	ORGANIS
NVP	N-VP	N-VINYLPHTHALIMIDE	COMP
NVZ		NEDERLANDSE VERENIGING VAN ZEEFABRIKANTEN (NL)	ORGANIS
NWG	LOD	NACHWEISGRENZE	EVAL

Abkürzung Akromym abbreviation akronym	Alternative alternative	Bedeutung meaning	Zuordnung related section
OAA	FAO	ORGANISATION DES NATIONS UNIES POUR L'ALIMENTATION ET L'AGRICULTURE	INSTI
OAD		ONE ATOM DETECTION	METH
OAQPS		OFFICE OF AIR QUALITY PLANNING AND STANDARDS	INSTI
OAS		OBERFLAECHENAKTIVES ADSORBENZ	MATER
OAS		OPTOACOUSTIC SPECTROMETRY	METH
OASH		OFFICE OF THE ASSISTANT SECRETARY FOR HEALTH (INT)	INSTI
OC		ON COLUMN TECHNIC	INSTR
OCAD		o-CHLOROBENZALDEHYDE	COMP
OCAMS		ORBITAL CORRESPONDENCE ANALYSIS IN MAXIMUM SYMMETRY	THEOR
OCBA		o-CHLOROBENZOIC ACID	COMP
OCBC		o-CHLOROBENZYL CHLORIDE	COMP
OCBN		o-CHLOROBENZONITRILE	COMP
OCCA		OIL AND COLOR CHEMISTS ASSOCIATION (US)	ORGANIS
OCCN		o-CHLOROBENZYL CYANIDE	COMP
OCDC		o-CHLORODICHLOROTOLUENE	COMP
OCDD	PCDD8	OCTACHLORO-p-DIBENZO DIOXIN	COMP
OCDE	OECD	ORGANISATION DE COOPERATION ET DE DEVELOPMENT ECONOMIQUE (PARIS-INT)	ORGANIS
OCDF	PCDF8	OCTACHLORO DIBENZO FURAN	COMP
OCF		OWENS CORNING FIBREGLAS	MATER
OCGC		ON COLUMN GASCHROMATOGRAPHY	METH
OCOC		o-CHLOROBENZOYL CHLORIDE	COMP
OCPA		o-CHLOROPHENYLACETIC ACID	COMP
OCPT		2-CHLORO-4-AMINOTOLUENE (ORTHO-CHLORO-PARA-AMINOTOLUENE)	COMP
OCR		OPTICAL CHARACTER RECOGNITION	EVAL TECHN
OCS		OCTACHLORO STYRENE	COMP
OCT		o-CHLOROTOLUENE	COMP
OCT		ORNITHINE CARBAMYL TRANSFERASE	CLINCHEM
OCTC		o-CHLOROBENZOTRICHLORIDE	COMP
OCTEO		OCTYLTRIETHOXYSILANE	COMP
OD	O.D.	OPTICAL DENSITY	INSTR
ODA		4,4'-OXYDIANILINE	COMP
ODEPA		N-(3-OXAPENTAMETHYLENE) N',N''-DIETHYLENE PHOSPHORAMIDE	COMP

Abkürzung Akromym abbreviation akronym	Alternative alternative	Bedeutung meaning	Zuordnung related section
ODMR		OPTICALLY DETECTED (DETERMINED) MAGNETIC RESONANCE SPECTROSCOPY	METH
ODPN		3,3'-OXYDIPROPIONITRIL	COMP
ODPN		BETA,BETA-OXYDIPROPIONITRIL	MATER
ODS		ORIENTATION DISTRIBUTION FUNCTION	THEOR
ODSC		OCTADECYL SILANE COLUMNS	INSTR
OE		OPTICAL EMISSION	METH
OECD		ORGANIZATION FOR ECONOMIC COOPERATION AND DEVELOPMENT (INT)	INSTI
OED	PSD	ORTSEMPFINDLICHER DETECTOR	INSTR
OEG		OLIGO ETHYLENE GLYCOL	MATER
OEGMM		OLIGO ETHYLENE GLYCOL MONO METHYL ETHER	MATER
OEM		ORIGINAL EQUIPMENT MANUFACTURER (INT)	ORGANIS
OEP		OCTAETHYLPORPHYRIN-LIGAND	COMP
OES	AES	OPTICAL EMISSION SPECTROSOPY	METH
OF		OXIDATION-FERMENTATION	CLINCHEM
OFCCP		OFFICE OF FEDERAL CONTRACT COMPLIANCE PROGRAMS (US)	INSTI
OFIMAT	BIGA	OFFICE FEDERAL DE L'INDUSTRIE,DES ARTS ET METIERS ET DU TRAVAIL (CH)	INSTI
OGV		OPTIMAL GAS VELOCITY	INSTR
OGY		OXYTETRACYCLINE-GLUCOSE YEAST EXTRACT	CLINCHEM
OHE		OFFICE OF HEALTH AND ECONOMICS	INSTI
OHEA		OFFICE OF HEALTH AND ENVIRONMENT ASSESSMENT (EPA,US)	INSTI
OHI		OCCUPATIONAL HEALTH INSTITUTE (US)	INSTI
OHM		OCTADECYL HYDROGEN MALEATE	COMP
OHMR		OFFICE OF HAZARDOUS MATERIALS REGULATIONS	INSTI
OICM	IKS	OFFICE INTERKANTONAL DE CONTROLE DES MEDICAMENTS (CH)	INSTI
OIML		ORGANISATION INTERNATIONAL DE METROLOGIE LEGAL (INT)	NORM ORGANIS
OIT	ILO	ORGANISATION INTERNATIONAL DU TRAVAIL (GENEVE ,INT)	INSTI
OMA		OPTICAL MULTICHANNEL ANALYSIS	METH
OMB		OFFICE OF MANGEMENT AND BUDGET (US)	INSTI
OMC		ORGANOMETALLIC CHEMISTRY	COMP

Abkürzung Akromym abbreviation akronym	Alternative alternative	Bedeutung meaning	Zuordnung related section
OMD	L-3-MTO	3-O-METHYLDOPA	COMP
OMH-1		SODIUM DIETHYLDIHYDROALUMINATE	COMP
OMI		o-METHYL-ISOUREA	COMP REAG
OMP		OROTIDINE-5'-MONOPHOSPHATE	COMP
OMP		OLIGO-N-METHYLMORPHOLINIUM PROPYLENE OXIDE	COMP
OMS	WHO	ORGANISATION MONDIAL SANTE (UN , INT)	INSTI
ON	ÖN	ÖSTEREICHISCHES NORMUNGSINSTITUT (Ö)	NORM
ONO		8-HYDROXYCHINOLIN-N-OXID	COMP
ONP		o-NITROBIPHENYL	COMP.
ONUDI	UNIDO	ORGANISATION DES NATIONS UNIES POUR LE DEVELOPMENT INDUSTRIEL (INT)	INSTI
OODR		OPTICAL OPTICAL DOUBLE RESONANCE SPECTROSCOPY	METH
OP	OP-Comp	ORGANIC PHOSPHOROUS COMPOUNDS	COMP
OPA		o-PHTHALALDEHYDE	COMP
OPAP		MOPA + DOPA	COMP
OPD		OCEAN DRILLING PROJECT (INT)	TECHN
OPEC		ORGANIZATION OF PETROLEUM EXPORTING COUNTRIES (INT)	ORGANIS
OPEPR		OPTICAL PERTUBATION ELECTRON PARAMAGNETIC RESONANCE	METH
OPG		OXY POLY GELANTINE	CLINCHEM
OPGV		OPTIMAL PRACTICAL GAS VELOCITY	INSTR
OPLC		OVER PRESSURE LAYER CHROMATOGRAPY	METH
OPPA		OCTYL PYROPHOSPHINOXIDE	COMP
OPPC		OPEN PARALLEL PLATE COLUMNS	INSTR
OPT		ORTHO-PHTHALALDEHYDETHIOETHANOL	COMP
OPTA		o-PHTHALDIALDEHYDE	COMP
OPTLC		OVER PRESSURE THIN LAYER CHROMATOGRAPHY	METH
OPTS		OFFICE OF PESTICIDES AND TOXIC SUBSTANCES (EPA,US)	INSTI
ORD		OPTICAL ROTATIONS DISPERSION	METH
ORD		OFFICE OF RESEARCH AND DEVELOPMENT (EPA,US)	INSTI
ORDIS		OFFICE OF RESEARCH AND DEVELOPMENT INFORMATION SYSTEM (EPA,US)	LIT
ORGALIME		ORGANISME DE LIAISON DES INDUSTRIES METALLIQUES EUROPEENNES (INT)	ORGANIS

Abkürzung Akromym	Alternative	Bedeutung	Zuordnung
abbreviation akronym	alternative	meaning	related section
ORM		OVERLAPPING RESOLUTION MAP (HLPC)	EVAL
ORM		OTHERWISE REGULATED MATERIALS (KENNZEICHNUNG)	COMP MATER
ORN	Orn	ORNITHINE	COMP
ORNL		OAK RIDGE NATIONAL LABORATORY (US)	INSTI
OROM	O-ROM	OPTICAL READ ONLY MEMORY	ELECT
ORS		ORTHOPHTHALALDEHYDE REACTIVE SUBSTANCES	COMP
ORTEP		TNO ORGANO TIN ENVIRONMENTAL PROJECT (UTRECHT,NL)	ENVIR
ORUF		OLEAT REPLACEMENT ULTRAFILTRATION	CLINCHEM
OSHA		OCCUPATIONAL SAFETY AND HEALTH ADMINISTRATION (US)	ORGANIS
OSIRIS	Osiris	OXIDISPER.SINGLECRYSTALS IMPROVED BY RESOLIDIFICATION IN SPACE	MATER TECHN
OST		OFFICE OF SCIENCE AND TECHNOLOGY (US)	INSTI
OSW		OFFICE OF SOLID WASTE (EPA,US)	INSTI
OTA		OFFICE OF TECHNOLOGY ASSESSMENT (US)	INSTI
OTB		o-TOLUIDINE BORIC ACID	COMP
OTC		ORNITHINE TRANSCARB AMYLASE	CLINCHEM
OTC		OVER THE COUNTER ARTICLES	PHARM
OTD		o-TOLUENEDIAMINE	COMP
OTS		OFFICE OF TOXIC SUBSTANCES (US)	INSTI
OTTLE		OPTICALLY TRANSPARENT THIN LAYER ELECTRODE	INSTR
OV		OHIO VALLEY SPECIALITY CHEMICAL CORPORATION (US)	COMPANY
OYE	NADPH2	OLD YELLOW ENZYME	CLINCHEM
ÖKOBUR		ARBEITSKREIS ÖKOLOGISCHE BEURTEILUNG (VCI,D)	ORGANIS

Abkürzung Akromym	Alternative	Bedeutung	Zuordnung
abbreviation akronym	alternative	meaning	related section
PA		POLYAMIDE &POLYALOMER	MATER
PA	pa	PRO ANALYSI	COMP
PA	PA-EFFECT	PHOTOACOUSTIC EFFECT	INSTR
PA		PHOSPHATIDYLINOSITOL	CLINCHEM
PA		PHOPHATIDIC ACID	CLINCHEM
PA		POSITRON ANNIHILATION	INSTR
PAA		PRIMARY AROMATIC AMINES	COMP
PAA		POLYCYCLIC AROMATIC AMINES	COMP
PAA		PHOTON ACTIVATION ANALYSIS	METH
PAB	PABA	4-AMINO-BENZOICACID =p-AMINOBENZOIC ACID	COMP
PABA	PAB	p-AMINOBENZOIC ACID	COMP
PAC		POLYACETYLENE	MATER
PAC	PCA	POLYCYCLIC AROMATIC COMPOUND	COMP
PAC		PERTURBED ANGULAR CORRELATION OF GAMMA-RAYS	METH
PAC		PIGMENTS ADVISORY COMMITTEE (ETAD,BASLE)	ORGANIS
PACEET		PROGRAMME ACTIVITY CENTRE FOR ENVIRONMENTAL EDUCATION AND TRAINING (UN)	INSTI
PACIA		PARTICLE AGGLUTINATION COUNTER IMMUNO ASSAY	METH
PAD		PULSED AMPEROMETRIC DETECTOR	INSTR
PAD		POST ACCELERATION DETECTOR	INSTR
PADA		POLY(ADIPIC ANHYDRIDE)	MATER
PADA		2-(5-BROMO-2-PYRIDYLAZO)-5-DIETHYLAMINOPHENOL	COMP
PADAP		2-(2-PYRIDYLAZO)-5-DIETHYLAMINOPHENOL	COMP
PADS		POTASSIUM PEROXYLAMINE DISULFONATE	COMP
PAES		PHOTON INDUCED AUGER ELECTRON SPECTROSCOPY	METH
PAFC		PHOSPHORIC ACID FUEL CELL	TECHN
PAG		POLYACRYLAMIDE GEL	MATER
PAGIF		POLYACRYLAMIDE ISO FOCUSSING	METH
PAH	PAK	POLYCYCLIC AROMATIC HYDROCARBON	COMP
PAH		p-AMINOHIPPURIC ACID	COMP
PAI		PESTICIDES ASSOCIATION OF INDIA (IND)	INSTI
PAK	PAH	POLYCYCLISCHE AROMATISCHE KOHLENWASSERSTOFFE	COMP

Abkürzung Akromym	Alternative	Bedeutung	Zuordnung
abbreviation akronym	alternative	meaning	related section
PAL		PHENYLALANINE AMMONIA LYASE	CLINCHEM
PAM		PROGRAM OF ACTIONS AND METHODS	EVAL
PAM	2-PAM	2-PYRIDINE ALDOXIME METHIODIDE (=PRALIDOXIMIODIDE)	COMP
PAM	L-PAM	L-PHENYLALANIE MUSTARD =MELPHALAN	COMP
PAMBA		p-AMINO-METHYL BENZOIC ACID	COMP
PAMCL	2-PAMCl	PYRIDINE-2-ALDOXIME METHOCHLORIDE	COMP
PAMF		PYRAZOLAM ARYLMETHAN FARBSTOFFE	MATER
PAMS		PRECISION ABRAISON MASS SPECTROMETRY	METH
PAN		PEROXYACETYLNITRATE	COMP
PAN		PHENYL-1-NAPHTHYLAMINE (=PHENYL ALPHA NAPHTHYLAMINE)	COMP
PANH		POLY AROMATIC NITROGEN HETEROCYCLES	COMP
PAP		PROSTATIC ACID PHOSPHATASE	CLINCHEM
PAP		PEROXIDATE ANTIBODY COMPLEX PEROXIDASE	CLINCHEM
PAP		O,O-DIMETHYL S-ALPHA-(ETHOXYCARBONYL)BENZYL PHOSPHOROTHIOLOTHIOATE	COMP
PAPA		POLY(AZELAIC ANHYDRIDE)	COMP
PAPI		POLYMETHYLENE POLYPHENYL ISOCYANATE	MATER
PAPP		1-(4-AMINOPHENYL)-1-PROPANONE 4=PARA	COMP
PAPS		3'-PHOSPHO ADENOSINE-5'PHOSPHOSULFATE	CLINCHEM
PAQH		PYRIDINE-2-CARBALDEHYD-2-QUINOLYHYDRAZONE	COMP
PAR		4-(2-PYRIDYLAZO) RESORCINOL	COMP
PARS		PHOTOACOUSTIC RAMAN SPECTROSCOPY	METH
PAS		p-AMINO-SALICYCLIC ACID	COMP
PAS		PHOTOACOUSTIC SPECTROSCOPY	METH
PAS		PERIODIC ACID - SCHIFF STAINING	CLINCHEM
PAS		POSITRON ANNIHILATION SPECTROSCOPY	METH
PASCA		POSITRON ANNIHILATION SPECTROSCOPY FOR CHEMICAL ANALYSIS	METH
PASEM		PARTICLE ANALYSIS SCANNING ELECTRON MICROSCOPY	METH
PASH		POLYAROMATIC SULFUR HETEROCYCLES	COMP
PAT		1-PHENYL-5-AMINOTETRAZOLE	COMP
PAZ	PAZ-	(4-PHENYLAZO)BENZYL FORMATE-LIGAND	COMP

Abkürzung Akromym	Alternative	Bedeutung	Zuordnung
abbreviation akronym	alternative	meaning	related section
PB		POLYBUTENE-1	MATER
PB		PROJECTILE BREMSTRAHLUNG	SPECT
PBA		p-BENZOQUINONE-2,3-DICARBOCYCLIC ACID	COMP
PBA		PHENYLBORONICACID - SILICA BONDED	MATER
PBB	PBB's	POLYBROMOBIPHENYL	COMP
PBBO		2-BIPHENYL-4-YL-6-BENZOXAZOLE	COMP
PBC		PRIMARY BILAR CIRRHOSE	CLINCHEM
PBD		2-(4-BIPHENYLYL)-5-PHENYL-1,3,4-OXADIAZOLE	COMP
PBE		PERMETHRINIC ACID PENTAFLUOROBENZYLESTER	COMP
PBG		PHORPHOBILINOGEN	CLINCHEM
PBI		p-BENZOQUINONE-2,3-DICARBOXYCLIC ACID	COMP
PBMC		PERIPHERE BLOOD MONO NUCLEOUS CELLS	CLINCHEM
PBN		PHENYL-TERT-BUTYLNITRIDE (N-TERT-BUTYL-ALPHA PHENYLNITRON)	COMP
PBN		PHENYL BETA NAPHTHYLAMINE (=PHENYL-2-NAPHTHYLAMINE)	COMP
PBN		PYROLYTIC BORON NITRIDE	MATER
PBP		PROTON BLOCKING PATTERN	METH
PBP		p-(BENZYLOXY)PHENOL	COMP
PBPB		p-BROMOPHENACYLBROMIDE	COMP
PBPH		(LITHIUM-PERHYDRA)-9-BORAPHENYL YL HYDRIDE	COMP
PBS		PHOSPHATE BUFFERED SOLUTE/SALINE	MATER
PBS		POLY(BUTENE-1-SULFONE)	MATER
PBT		POLYBUTYLENE TEREPHTHALATE	MATER POLYM
PBX		POTASSIUM N-BUTYL XANTHATE	COMP
PC		PERSONAL COMPUTER	INSTR
PC		POLYCARBONATE	MATER POLYM
PC	Pc-	PHTHALOCYNANINE (LIGAND)	COMP
PC		PAPER CHROMATOGRAPHY	METH
PC		PHOSPHATIDYLCHOLIN	CLINCHEM
PCA		PARIETOL-CELL-ANTIBODY	CLINCHEM
PCA		PENTACHLOROANILINE	COMP
PCA		PRINCIPAL COMPONENT ANALYSIS	EVAL
PCA	PAC	POLYCYCLIC AROMATIC COMPOUND	COMP
PCA		PACKAGING COUNCIL OF AUSTRALIA (AUS)	INSTI
PCAB		POLYCHLORO-ALKYLBIPHENYL	COMP

Abkürzung Akromym	Alternative	Bedeutung	Zuordnung
abbreviation akronym	alternative	meaning	related section
PCAD		p-CHLOROBENZALDEHYDE	COMP
PCB	PCB's	POLYCHLOROBIPHENYL	COMP
PCB		PRIMAER CHRONISCHE POLYARTHRITIS	CLINCHEM
PCBA		p-CHLOROBENZOIC ACID	COMP
PCBC		p-CHLOROBENZYL CHLORIDE	COMP
PCBN		p-CHLOROBENZONITRILE	COMP
PCBTF		p-CHLOROBENZOTRIFLUORIDE	COMP
PCBZ	PCBz n	POLYCHLOROBENZENE (n=2,3,4,5)	COMP
PCC	PCC's	POLYCHLORO CAMPHENE	COMP
PCC		PYRIDINIUMCHLORO CHROMATE	COMP
PCCH		PENTACHLORO CYCLOHEXANE	COMP
PCCN		p-CHLOROBENZYL CYANIDE	COMP
PCDC		p-CHLORODICHLOROTOLUENE	COMP
PCDD	PCDD(X)	POLYCHLORO-p-DIBENZO DIOXINE (X= 2 - 8)	COMP
PCDE		POLYCHLORO DIPHENYLETHER	COMP
PCDF	PCDF(x)	POLYCHLORO DIBENZO FURAN (x=2 - 8)	COMP
PCDPE(s)	PCDE	POLYCHLORO DIPHENYLETHER(S)	COMP
PCEM		PHASE CONTRAST ELECTRON MICROSCOPY	METH
PCF		PAIR CORRELATION FUNCTION	THEOR
PCH		TRANS-4-PROPYL-(4-CYANOPHENYL)-CYCLOHEXANE	COMP
PCH		PARENT COMPOUND HANDBOOK (LIT, CA)	LIT
PCH2	PcH2	PHTHALOCYANINE	COMP
PCI		POSITIVE CHEMICAL IONIZATION	INSTR
PCILO		PERTUBATIVE CONFIGURATION INTERACTION OVER LOCALIZED ORBITALS	THEOR
PCLC		PREPARATIVE COLUMN LIQUID CHROMATOGRAPHY	METH
PCM		PULSE CODE MODULATION	INSTR
PCM		PULSE COUNTING METHOD	METH
PCM		PHASE CHANGE MATERIAL	MATER
PCMB		p-CHLOROMERCURIBENZIOC ACID	COMP
PCMX		p-CHLORO-m-XYLENOL	COMP
PCN	PxCN	POLYCHLORO NAPHTHALENE (x= 2 - 8)	COMP
PCNB		PENTACHLORO NITROBENZENE	COMP
PCOC		p-CHLOROBENZOYLCHLORIDE	COMP
PCONA		p-CHLORO-o-NITROANILINE	COMP
PCOT		p-CHLORO-o-AMINOTOLUENE =4-CHLORO-2-AMINOTOLUENE	COMP

Abkürzung Akromym	Alternative	Bedeutung	Zuordnung
abbreviation akronym	alternative	meaning	related section
PCP		PENTACHLOROPHENOL	COMP
PCP		PHENACYCYCLIDINE	COMP CLINCHEM
PCPA		p-CHLOROPHENOXYACETIC ACID =4-CHLOROPHENOXYACETIC ACID	COMP
PCPA		p-CHLOROPHENYL-ALANINE	COMP
PCPA		PENTACHLOROANISOLE	COMP
PCPH	PCPh	POLYCHLOROPHENOL	COMP
PCPP(s)		POLY CHLORINATED PHENOXYPHENOLS	COMP
PCR		POST COLUMN REACTION	METH
PCRS		POST COLUMN REACTION SYSTEM	INSTR
PCST	PCSt	POLYCHLOROSTYRENE	COMP
PCT		POLYCHLORO TERPHENYL	COMP
PCT	PCT-	PHENYL-THIOCARBONYL-LIGAND	COMP
PCT		PATENT COOPERATION TREATY (EUR)	REG
PCT		p-CHLOROTOLUENE	COMP
PCT		PORPHYRIA CUTANEA TARDA	CLINCHEM
PCTA		PENTACHLORO THIOANISOLE (=PENTACHLORO-THIOMETHYL-ANILINE)	COMP
PCTC		p-CHLOROTRICHLOROTOLUENE	COMP
PCTF		PHASE CONTRAST TRANSFER FUNCTION	THEOR
PCTFE		POLYCHLORO TRIFLOUROETHYLENE	MATER POLYM
PD	PD50	PROPHYLACTISCHE DOSIS	CLINCHEM
PD		PHOTO DESORPTION	METH
PD	TE	PARTIAL DISCHARGE	TECHN
PDA		p-PHENYLENE DIAMINE	COMP
PDA		PHORBOL 12,13-DIACETATE	COMP
PDA		PULSE HIGHT DISTRIBUTION ANALYSIS	EVAL METH
PDA	DAD	PHOTHODIODE ARRAY	INSTR
PDB		p-DICHLORO BENZENE	COMP
PDBZ	PDBz	PHORBOL 12,13-DIBENZOATE	COMP
PDC		PYRIDINIUM DICHROMATE	COMP
PDCA		PYRIDINE 2,6-DICARBOXYLIC ACID	COMP
PDD		p-DICHLOROBENZENE	COMP
PDDCC		PLANT DISEASE DIVISION CULTURE COLLECTION AUCKLAND (NZ)	INSTI
PDE		3',5'-NUCLEOTIDE PHOSPHODIESTERASE	CLINCHEM
PDE		PARTIAL DIFFERENTIAL EQUATION	THEOR
PDEA		PHENYL DIETHANOLAMINE	COMP

Abkürzung Akromym abbreviation akronym	Alternative alternative	Bedeutung meaning	Zuordnung related section
PDEAS		PHENYLDIETHANOLAMINE SUCCINATE	COMP
PDF		POWDER DIFFRACTION FILE	METH
PDGF		PLATELET-DERIVED GROWTH FACTOR	CLINCHEM
PDL		PULSE DYE LASER	INSTR
PDMEA	P-DMEA	2-DIMETHYLAMINOETAHANOL-DIHYDOGENPHOSPHATE	COMP
PDMS		PLASMA DESORPTION MASS SPECTROSCOPY	METH
PDQ		SODIUM (2-METHYL-4-CHLOROPHENOXY)BUTYRATE	COMP
PDR		PHARMADOKUMENTATIONRING (PBI,D)	LIT ORGANIS
PDS	2-PDS	2,2'-DITHIODIPYRIDINE	COMP
PDSC		PRESSURE DIFFERENTIAL SCANNING CALORIMETRY	METH
PDT		PHOTODYNAMIC THERAPY	CLINCHEM
PDT		3-(2-PYRIDYL)-5,6 DIPHENYL-1,2,4 TRIAZINE	COMP REAG
PE		POLYETHYLENE	MATER POLYM
PE	PEA	PHOSPHATIDYLETHANOLAMINE	COMP
PE		PERKIN ELMER-BODENSEEWERK	COMPANY
PEA	PE	PHOSPHATIDYLETHANOLAMINE	CLINCHEM
PEA		N-PHENYL ETHANOLAMINE =N-(2 HYDROXYETHYL)ANILINE	COMP
PEBG	PEDG	1-PHENETYL BIGUANIDE	COMP
PEC	4-PEC	S-(2-(4-PYRIDYL)ETHYL)L-CYSTEIN)	COMP
PEEA		N-PHENYL-N-ETHYL ETHANOLAMINE = N(2-HYDROXYETHYL)-N-ETHYLANILINE	COMP
PEEE		PHOTOSTIMULATED EXO-ELECTRON EMISSION	METH
PEEK		POLYETHYLENE ETHERKETON	COMP
PEEM		PHOSPHORESCENCE EMISSION EXCITATION MATRIX	INSTR
PEEP		POSITIVE END EXSPIRATONICAL PRESSURE	CLINCHEM
PEF		PROJ. EUROP FORSCHUNGSZENTRUM F LUFTREINHALTUNG	ENVIR
PEGDM		POLYETHYLENE GLYCOL DIMETHYL ETHER	MATER
PEHA		PENTAETHYLENEHEXAMINE	COMP
PEHD		POLYETHYLENE HIGH DENSITY	MATER
PEI		POLYETHERIMIDE	MATER
PEI-C		POLYETHYLENEIMINE INTEGRATED CELLULOSE	MATER

Abkürzung Akromym abbreviation akronym	Alternative alternative	Bedeutung meaning	Zuordnung related section
PEL		PERMISSIBLE EXPOSURE LEVELS OR LIMITS (OSHA, US)	REG
PELS		PROPIONATE ERYTHROMYCIN LAURYL SULFATE	COMP
PEM		PIEZOELASTIC MODULATION	INSTR
PEM		MONOMETHYL PHOSPHATIDYLETHANOLAMINE	CLINCHEM
PEM		PHOTO ELECTRON MICROSCOPY	METH
PEMA		2-ETHYL-2-PHENYLMALONAMIDE	COMP
PEMM		DIMETHYLPHOSTPHATIDYLETHANOLAMINE	COMP
PENIS		PROTON ENHANCED NUCLEAR INDUCTION SPECTROSCOPY	METH
PEO		POLYETHYLENE OXIDE	MATER
PEP		POLYEPOXID	MATER POLYM
PEP	4-PEP	S-(2-(4-PYRIDYL)ETHYL)-DL-PENICILLAMINE	COMP
PEP		PHOSPHOENOLPYRUVIC ACID	CLINCHEM COMP
PEP	PEP-	2-PHENYL-(ETHYLPHENANTHIDINIUM)-LIGAND	COMP
PEP	PEP-Na	PHOSPHOENOL PYRUVAT	COMP
PES		PHOTOELECTRON SPECTROSCOPY	METH
PESIS		PHOTO ELECTRON SPECTROSCOPY OF INNER SHELL(S)	METH
PESM		PHOTOELECTRON SPECTRAL MICROSCOPY	METH
PESOS		PHOTO ELECTRON SPECTROSCOPY OF OUTER SHELL(S)	METH
PET		PHOTO EMISSION TOMOGRAPHY	CLINCHEM
PET		PENTACRYLTHRITOL	MATER
PET		POLY(ETHYLENE TERPHTHALATE)	MATER
PET		POSITRON EMISSION TOMOGRAPHY	CLINCHEM
PETA		PENTAERYTHRITOL TRIACRYLATE	COMP
PETN		PENTA ERYTHRITOL TETRANITRATE	COMP
PETP		POLYETHYLENE TEREPHTHALATE	COMP
PETTCA		PIGMENT ECOLOG.& TOXICOLOGICAL TECHNICAL COMMITTE OF AUSTRALIA (AUS)	INSTI
PEX		PRESSURE EXPANSION DIGESTION	INSTR
PF		PHENOL-FORMALDEHYDE RESIN	MATER POLYM
PFA		PERFLUOROALKANE	COMP
PFA		PERFLUOROETHYLENE ALKYLETHER COPOLYMER	MATER
PFAA		PERFLUOROACETIC ACID ANHYDRIDE	COMP

Abkürzung Akromym	Alternative	Bedeutung	Zuordnung
abbreviation akronym	alternative	meaning	related section
PFB	PFB-	PENTAFLUOROBENZOYL - LIGAND	COMP
PFBHA		PENTAFLUORO BENZALDEHYDE	COMP
PFCP		PERFLUOROCYCLOPENTENE	COMP
PFE		PHOSPHATED FATTY ALCOHOL ETHOXYLATE	MATER
PFEP		POLYFLUOROISOBUTYLENE	MATER
PFG		PULSED FIELD GEL ELECTROPHORESIS	METH
PFK		PERFLUOROKEROSANE	COMP
PFLSCHG	PflSchG	PFLANZENSCHUTZGESETZ (D)	REG
PFM		PULSE FREQUNCE MODULATION	INSTR
PFMS		PYROLYSIS FIELD IONIZATION MASS SPECTROMETRY	METH
PFP	PFP-	PENTAFLUOROPROPYL-LIGAND	COMP
PFPA		PERFLUORO PROPIONIC ANHYDRIDE	COMP
PFPI		PENTAFLUOROPROPIONYLIMIDAZOLE	COMP REAG
PFR		PLUG-FLOW REACTOR	TECHN
PFTBA		PERFLUORO-TRI-N-BUTYLAMINE	COMP
PG		PHLOROGLUCINOL	COMP
PG		PYROLYTHIC GRAPHITE	MATER
PG		PHOSPHATIDYLGLYCEROL	CLICHEM
PG		PROSTAGLANDINE (DIV. PGD2,PGE2,PGF1ALPHA,PGF2ALPHA)	CLINCHEM COMP
PG		PROTECTIVE GROUP	COMP
PGB	PGB-	PROPYLENE GLYCOBUTYLETHER-LIGAND	COMP
PGC		PYROLISIS GASCHROMATOGRAPHY	METH
PGC		PROCESS-GASCHROMATOGRAPHY	TECHN
PGE		PROSTAGLANDIN E	CLINCHEM
PGI2		PROSTACYCLINE	CLINCHEM
PGK		PHOSPHOGLYCERAT KINASE	CLINCHEM
PGL		POLYGLUTARALDEHYDE	COMP
PGM		POLYETHYLENEGLYCOLDIMETHACRYLATE	MATER
PGNAA		PROMT GAMMA NEUTRON ACTIVATION ANALYSIS	METH
P2S		2-PYRIDINEALDOXIME METHYL METHANE SULFONATE	COMP
PH.BELG.	Ph.Belg.	PHARMACOPEE BELGE	NORM
PH.DAN.	Ph.Dan.	PHARMACOPOEA DANICA	NORM
PH.EUR.	Ph.Eur.	EUROPAEISCHE PHARMAKOPOE	NORM
PH.F.	Ph.F.	PHARMACOPOEA FENNICA	NORM
PH.FRANC.	Ph.Franc.	PHARMACOPEE FRANCAISE	NORM

Abkürzung Akromym abbreviation akronym	Alternative alternative	Bedeutung meaning	Zuordnung related section
PH.HELV.	Ph.Helv.	PHARMACOPOEA HELVETICA	NORM
PH.HISP.	Ph.Hisp.	PHARMACOPEA OFICIAL ESPANOLA	NORM
PH.INTERNA	Ph.Intern	PHARMACOPOEA INTERNATIONALIS	NORM
PH.NED.	Ph.Ned.	PHARMACOPOEA NEDERLANDICA	NORM
PH.NORD.	Ph.Nord.	PHARMACOPOEA NORDICA	NORM
PH.NORV.	Ph.Norv.	PHARMACOPOEA NORVEGICA	NORM
PH.PORT.	Ph.Port.	PHARMACOPEIA PORTUGUESA	NORM
PH.SVEC.	Ph.Svec.	PHARMACOPOEA SVECICA	NORM
PHA		PHYTO HAEMACY GLUTININE	CLINCHEM
PHB		p-HYDROXYBENZOICACID	COMP
PHB		PROYL HYDROXY BENZOATE	COMP
PHD		PLASMA HYDRIDE DEVICE	INSTR
PHE	Phe	PHENYLANALINE	COMP
PHEEM		PHOTO EMISSIONS ELECTRON MICROSCOPY	METH
PHG	Phg	PHENYLGLYCINE	COMP
PHGA		PTEROYL HEPTAGLUTAMIC ACID	COMP
PHI		PHOSPHAHEXASE-ISOMERASE	CLINCHEM
PHI		PARAT HORMON INHIBITORS	CLINCHEM
PHR		PEAK HEIGHT RATIO	EVAL
PHS		PUBLIC HEALTH SERVICE (US)	INSTI
PHT		PYRROLIDONE HYDRO TRIBROMIDE	COMP
PHT		N-ETHOXY CARBONYL PHTHALIMIDE	COMP
PI		PHTHALIMIDE	COMP
PI		PHOSPHATITYLINOSITOL	COMP
PIA		PHENYLIODOSO DIACETATE	COMP
PIA		ARBEITSGEM. PRAKTISCHE INSTRUMENTELLE ANALYTIK	ORGANIS
PIA		PHARMACEUTICAL INDUSTRIES ASSOCIATION (CH)	ORGANIS
PIB		POLY ISOBUTYLENE	MATER POLYM
PIC		PHARMACEUTICAL INSPECTIONS CONVENTION	REG
PIC		PHENYL ISOCYANATE	COMP
PIC		PAIRED ION CHROMATOGRAPHY	METH
PICI		POSITIVE ION CHEMICAL IONIZATION	INSTR
PICT		PHENYL ISOTHIOCYANATE	COMP
PID		PHOTO ION DETECTOR	INSTR
PID		PROPORTIONAL-INTEGRAL-DIFFERENTIAL (DERIVATIVE)	TECHN

Abkürzung Akromym	Alternative	Bedeutung	Zuordnung
abbreviation akronym	alternative	meaning	related section
PIDFTMS	PID-FTMS	PARTICLE INDUCED DESORPTION FOURIER TRANSFORM MASS SPECTROMETRY	METH
PIH		PROLACTINE INHIBITING HORMONE	CLINCHEM
PIM		PEAK INTEGRATION METHOD	METH
PIMA		PRINTING INK MANUFACTURERS ASSOCIATION OF AUSTRALIA (AUS)	ORGANIS
PIO		PARALLEL INPUT-OUTPUT CONTROLLER	ELECT
PIP		PHOSPHATIDYL INOSITOL PHOSPHATE	COMP
PIPES		PIPERAZINE-1,4-DIETHANE-SULFONIC ACID	COMP
PIR		PHOTON INDUCED X-RAY	METH
PIS		PENNING IONIZATION SPECTROSCOPY	METH
PITC		PHENYL ISOTHIOCYANATE	COMP
PIX		PARTICLE INDUCED X-RAYS	INSTR
PIXE		PARTICLE (PROTON) INDUCED X-RAY EMMISSION	METH
PIXES		PARTICLE INDUCED X-RAY EMISION SPECTROSCOPY	METH
PKL		PLAST- OCH KEMIKALIELEVERANTÖRES FÖRENING (S)	ORGANIS
PKN		PRODUKTKONTROLNÄMNDEN (S)	INSTI
PLAP		PULSED LASER ATOMPROBE TECHNIQUE	METH
PLB		POROUS LAYER BEADS	MATER
PLC		PLASMA CHROMATOGRAPHY	METH
PLC		PROCESS LIQUID CHROMATOGRAPHY	METH
PLC	PSC	PREPARATIVE LAYER CHROMATOGRAPHY	METH
PLEP		PHOSPHOLIPID-EXCHANGE PROTEIN	CLINCHEM
PLM		POLARIZED LIGHT MICROSCOPY	METH
PLOT		POROUS LAYER OPEN TUBULAR COLUMN	INSTR
PLP		PYRIDOXALPHOSPHATE	COMP
PLP		PROTEO LIPID	CLINCHEM
PLRV		POTATO LEAFRALL VIRUS	CLINCHEM
PLS		PARTIAL LEAST QUARES	EVAL
PLZT		BLEI-LANTHAN-ZIRKONAT-TITANAT KERAMIK	TECHN
PM		PHOTO MULTIPLIER	INSTR
PM		PENSKY-MARTENS (FLAMMPUNKT)	INSTR
PM		PHOSPHATIDYLMETHANOL	CLINCHEM
PMA		PLANE MIRROR ANALYZER	INSTR
PMA		PHORBOL 12-MYRISTATE 13-ACETATE	COMP
PMA	PMAC	PHENYL MERCURY ACETATE	COMP

Abkürzung Akromym	Alternative	Bedeutung	Zuordnung
abbreviation akronym	alternative	meaning	related section
PMA		POLYURETHANE MANUFACTURERS ASSOCIATION (US)	ORGANIS
PMA		PIGMENT MANUFACTURERS ASSOCIATION (AUS)	ORGANIS
PMDTA		PENTAMETHYLDIETHYLENETRIAMINE	COMP
PMEA		N-(2-HYDROXYETHYL)-N-METHYLANILINE	COMP
PMFG		PULSED MAGNETIC FIELD GRADIENT	SPECT
PMFS		PHENYLMETHYLSULFURYLFLUORIDE	COMP
PMH		PHENYLMERCURIC HYDROXIDE	COMP
PMHC		PENTAMETHYL HYDROXY CHROMAN	COMP
PMHS		POLYMETHYLHYDROGEN SILOXANE	MATER
PMI		3-PHENYL-5-METHYLISOXAZOLE	COMP
PMI-ACID		3-PHENYL-5-METHYLISOXAZOLE-CARBOXYLIC ACID	COMP
PMMA		POLY (METHYLACRYLATE) (= POLYMETHYL METHACRYLATE)	MATER
PMN		PREMANUFACTURED NOTIFICATION	ELECT INSTR
PMO		PERTURATIONAL MOLECULAR ORBITAL	THEOR
PMP		1,2,2,6,6-PENTAMETHYLPIPERIDINE	COMP
PMP		POLY METHYLPENTENE	MATER POLY
PMP		O,O-DIMETHYL S-(PHTHALIMIDOMETHYL)-PHOSPHORODITHIONATE	COMP
PMPE	PMPE-5RING	POLY-m-PHENYLETHER -5-RING	MATER
PMR	NMR,H-NMR	PROTON MAGNETIC RESONANCE	METH
PMS		N-METHYLPHENAZONIUM METHOSULFATE (=PHENAZINE METHOSULFATE)	COMP
PMS		PREGNANT MAN'S SERUM	CLINCHEM
PMSF		PHENYL METHYL SULFONYL FLUORIDE	COMP
PMSG		PREGNANT MAN SERUM GONATROPIN	CLINCHEM
PMT		PHOTOMULTIPLIER TUBE	INSTR
PNA		PROVISIONAL NOMENCLATURE APPENDIX (IUPAC,INT)	LIT
PNA	PNA(s)	POLY NUCLEAR AROMATICS	COMP
PNA		PRENEOPLASTIC ANTIGEN	CLINCHEM
PNASA		p-NITROANILINE-o-SULFONIC ACID	COMP
PNBA		p-NITROBENZYLOXYAMINE	COMP REAG
PNBDI		p-NITROBENZYL-N,N-DIISOPROPYLTHIOUREA	COMP REAG
PNBPA		p-NITROBENZYL-N-n-PROPYLAMINE	COMP
PNDO		PARTIAL NEGLECT OF DIFFERENTIAL OVERLAP	THEOR

Abkürzung Akromym abbreviation akronym	Alternative alternative	Bedeutung meaning	Zuordnung related section
PNE		PHOSPHATED NONYL PHENYL ETHOXYLATE	COMP
PNK		POLYNUCLEOTIDE KINASE	CLINCHEM
PNMR		PROTON/PULSED NUCLEAR MAGNETIC RESONANCE	METH
PNMT		PHENYLETHANOLAMINE-N-METHYL-TRANSFERASE	CLINCHEM
PNO		PAIR NATURAL ORBITALS	THEOR
PNOT		p-NITRO-o-TOLUIDINE	COMP
PNP		p-NITROPHENOL (= 4-NITROPHENOL)	COMP
PNPDPP		p-NITROPHENYL DIPHENYL PHOSPHATE	COMP
PNPG		ALPHA-p-NITROPHENYL GLYCERINE	COMP
PNPP		p-NITROPHENYL PHOSPHATE	COMP
PNUE	UNEP	PROGRAMME DES NATIONS UNIES POUR L'ENVIRONEMENT (UN)	ENVIR
PO	p.o.	PER OS	CLINCHEM
POBN	4-POBN	ALPHA -4-PYRIDYL-N-TERT-BUTYLNITRONE ALPHA-1-OXIDE	COMP
POC		PARTICULATE ORGANIC CARBON	MATER
POC		CYCLOPENTYL OXY CARBONYL	COMP
POCL		PHOTOXYGENATION CHEMILUMINESCENCE	METH
POD		PEROXIDASE	CLINCHEM
POE		POLYOXYETHYLENE	MATER
POF		PULSED OPTICAL FEEDBACK	INSTR
POM		POLYOXY METHYLENE	MATER
POM		PARTICULATE ORGANIC MATTER	COMP
POMC		PRO-OPIO MELANOCORTIN	CLINCHEM
POPOP		2,2'-p-PHENYLENE BIS (5-PHENYLOXAZOLE)	COMP
POPSO		PIPERAZIN-1,4-BIS-(2-HYDROXY-PROPANSULFANSAEURE)	COMP
POX		PURGEABLE ORGANIC HALOGENS	COMP
PP		PROTOPORPHYRIN	CLINCHEM
PP		POLYPROPYLENE	MATER POLYM
PPA		POLYPHOSPHORIC ACID	COMP
PPACK		D-PHENYLALANYL-PROLYL-ARGININE CHLOROMETHYL KETONE	COMP
PPB	ppb	PARTS PER BILLION (1:10 E 9)	EVAL
PPB	PPB's	POROUS POLYMER BEADS	MATER
PPB		POLYMERIC 1,4-DIPHENYL-1,3-BUTADIENE	MATER POLYM
PPD		2,5-DIPHENYL-1,3,4-OXADIAZOLE	COMP

Abkürzung Akromym abbreviation akronym	Alternative alternative	Bedeutung meaning	Zuordnung related section
PPDA		PHENYL PHOSPHORODIAMIDATE	COMP
PPDP		p-p'-DIPHENOL	COMP
PPE		POLYPHOSPHATE ESTER	COMP
PPINICI		PULSED POSITIVE ION NEGATIVE ION CHEMICAL IONIZATION	METH
PPL		POLYPYRROL	MATER
PPL	NCPPB	PLANT PATHOLOGY LABORATORY HARPENDEN (GB)	INSTI
PPM	ppm	PARTS PER MILLION (1 : 10 E 6)	EVAL
PPM		(2S,4S)-4 DIPHENYLPHOSPHINO-2-DIPHENYLPHOSPHINO-PYRROLIDINE	COMP
PPMS		POLY-p-METHYLSTYRENE	MATER POLYM
PPNCI		BIS(TRIPHENYLPHOSPHORANYLIDENE)-AMMONIUM CHLORIDE	COMP
PPO		2,5-DIPHENYLOXAZOLE	COMP
PPO		POLYPHENYLENE OXIDE	MATER POLYM
PPP	PPP-VERF.	METHOD TO PARISER,PARR ,POPLE	THEOR
PPP	3-PPP	3-(3-HYDROXY-PHENYL)-N-PROPYL-PIPERIDINE	COMP
PPQ	ppq	PARTS PER QUADRILLION (1 : 10 E 15)	EVAL
PPSE		1-TRIMETHYLSILYLPOLYPHOSPHATE	COMP REAG
PPSEB	PPSeb	POLYPROPYLENE SEBACATE	MATER
PPT	ppt	PARTS PER TRILLION (1 : 10 E 12)	EVAL
PPTS		PYRIDINIUM-p-TOLUENESULFONATE (=PYRIDINIUM TOLUENE-SULFONATE)	COMP
PR		PHENOL ROT	COMP
PRA		PROMT RADIATION ANALYSIS	METH
PRH		LTH-RELEASING HORMONE	CLINCHEM
PRIM		POLYMER REINFORCED REACTION INJECTION MOULDING	TECHN
PRO	Pro	PROLINE	COMP
PRO		PROTRIPTYLINE	COMP
PROD		PARAMAGNETIC RESONANCE BY OPTICAL DETECTION	METH
PROM		PROGRAMMABLE READ ONLY MEMORY	ELECT
PROXYL		2,2,5,5-TETRAMETHYL-1-PYRROLIDIN-3-YLOXY	COMP
PS		POLYSTYRENE	MATER POLYM
PS		PHOSPHATIDYLSERINE	CLINCHEM
PS-CL	PS-Cl	2-PYRIDINESULFENYL CHLORIDE	COMP

Abkürzung Akromym / abbreviation akronym	Alternative / alternative	Bedeutung / meaning	Zuordnung / related section
PSA		POTENTIOMETRIC STRIPPING ANALYZER / ANALYSIS	INSTR METH
PSA		PROSTATA SPECIFIC ANTIGEN	CLINCHEM
PSAC		PRESIDENT'S SCIENCE ADVISORY COMMITTEE (US)	INSTI
PSBA		POROUS STYRYL BORONIC ACID	MATER
PSC		PROGRAMMABLE SAMPLE CHANGER	INSTR
PSC	PLC	PRAEPARATIVE SCHICHT CHROMATOGRAPHIE	METH
PSD		PROGRAMMABLE DISPENSER	INSTR
PSD	PSPD	POSITION/PHASE SENSITIVE PROPORTIONAL DETECTOR	INSTR
PSE		PERIODIC SYSTEM OF THE ELEMENTS	COMP THEOR
PSE		PROJECT SICHERHEITSSTUDIEN ENTSORGUNG (D)	TECHN ENVIR
PSEE		PHOTO STIMULATED EXOELECTRON EMISSION	METH
PSF		POINT SPREAD FUNCTION	THEOR
PSF		POLYSULFON	MATER POLYM
PSG		PULSE SEQUENCE GENERATION	INSTR
PSI	psi	POUNDS PER SQUARE INCH	INSTR
PSIA	psia	POUNDS PER SQUARE INCH ABSOLUTE	INSTR
PSIG	psig	POUNDS PER SQUARE INCH GAUGE	INSTR
PSK		POLYSACCHARIDE K	MATER
PSPA		POLY(SEBACIC ANHYDRIDE)	COMP
PSPC	PSPD	POSITION SENSITIVE PROPORTIONAL COUNTER	INSTR
PSPD	PSPC	POSITION SENSITIVE PROPORTIONAL DETECTOR	INSTR
PSSD		POSITION SENSITIVE SCINTILLATION DETECTOR	INSTR
PSTV		POTATO SPINDLE TUBEN VIROID	CLINCHEM
PT		PROTHROMBIN TIME	CLINCHEM
PTAD		4-PHENYL-1,2,4-TRIAZOLINE-3,5-DIONE	COMP
PTAP		PHENYLTRIMETHYL AMMONIUM PERBROMIDE	COMP
PTB		PHYSIKALISCH TECHNISCHE BUNDESANSTALT (BRAUNSCHWEIG,D)	INSTI
PTBBA		p-TERT-BUTYLBENZOIC ACID	COMP
PTC		PHASE TRANSFER CATALYSIS	METH

Abkürzung Akromym abbreviation akronym	Alternative alternative	Bedeutung meaning	Zuordnung related section
PTC		PLASMA THROMBOPLASTIN COMPONENT	CLINCHEM
PTC	PITC	PHENYLISOTHIOCYANATE	COMP
PTC	PTC-	PHENYL THIOCARBAMYL-LIGAND	COMP
PTFE		POLYTETRAFLUOROETHYLENE	MATER POLYM
PTGA		PTEROYL TRIGLUTAMIC ACID	COMP
PTH		PHENYLTHIOHYDANTOIN	COMP REAG
PTH		PARATHYROID HORMONE (=PARAT HORMONE)	COMP
PTH		PROTYHIONAMIDE	COMP CLINCHEM
PTI		POTENTIAL TOXICITY INDEX	BIOS NORM
PTMO		n-PROPYLTRIMETHOXIMETHANE	COMP
PTMT		TERAMETHYLENE TEREPHTHALATE	COMP
PTO		PATENT AND TRADEMARK OFFICE (US)	INSTI
PTP		POLYTEREPHTHALATE	MATER POLYM
PTSA		p-TOLUENE SULFONIC ACID	COMP
PTSED		PESTICIDE AND TOXIC SUBSTANCES ENFORCEMENT DIVISION (EPA,US)	INSTI
PTSI		p-TOLUENESULFONYL ISOCYANATE	COMP
PTT		PARTIAL THROMBOPLASTIN TIME	CLINCHEM
PTV		PROGRAMMED TEMPERATURE VAPORIZER	INSTR
PTZ		PARTIELLE THROMBOPLASTINZEIT	CLINCHEM
PUF		POLYURETHANE FOAM	MATER
PUFA		POLYUNSATURATED FATTY ACIDS	COMP
PUR		POLYURETHANE	MATER POLYM
PUREX		PLUTONIUM-URANIUM RECOVERY BY EXTRACTION	TECHN
PVA		POLYVINYLALCOHOL	MATER
PVA		POTATO VIRUS A	CLINCHEM
PVAC		POLYVINYLACETATE	MATER POLYM
PVC		POLYVINYLCHLORIDE	MATER
PVD		PHYSICAL VAPOUR DEPOSITION	TECHN
PVDC		POLYVINYLIDENECHLORIDE	MATER POLYM
PVDF		POLYVINYLIDENEFLUORIDE	MATER POLYM
PVP		POLY-2-OXO-1-VINYL-PYRROLIDINE = POLYVINYLPYRROLIDONE	CLINCHEM MATER
PVP-I		POLYVINYLPYRROLIDONE-IODINE	MATER
PVPCC		POLYVINYL PYRIDINIUM CHLOROCHROMAT	MATER REAG
PVPDC		POLY(4-VINYLPYRIDINIUM)DICHROMATE	MATER

Abkürzung Akromym abbreviation akronym	Alternative alternative	Bedeutung meaning	Zuordnung related section
PVSK		POTASSIUM POLYVINYL SULFATE	COMP
PVT		PRESSURE-VOLUME-TEMPERATURE RELATION	EVAL
PVT		PULSED VIDEO THERMOGRAPHY	METH
PXA		PRIMARY X-RAY ANALYSIS	METH
PZA		PYRAZINAMIDE	COMP
PZT		PIEZOELECTRIC TRANSDUCER	INSTR

Abkürzung Akromym abbreviation akronym	Alternative alternative	Bedeutung meaning	Zuordnung related section
QAE	QAE-	QUARTARY AMINOETHYL GROUP	COMP
QC		QUALITY CONTROL	METH
QCL		QUASI CLASSICAL	THEOR
QCM		QUARTZ CRYSTAL MICROBALANCE	INSTR
QDE		QUANTUM DETECTION EFFICIENCY	INSTR
QELS		QUASI ELECTRIC LIGHT SCATTERING	METH
QHE		QUANTUM HALL EFFECT	EVAL
QM		QUANTITATIVE MICROSCOPY	METH
QMAS		QUADRUPOLE MASS ANALYSER FOR SOLIDS	INSTR
QMF		QUERSTROM MIKROFILTRATION	TECHN
QMS		QUADRUPOLE MASS SPECTROMETER	INSTR
QSAR	QSWB	QUANTITATIVE STRUCURE ACTIVITY RELATION	CLINCHEM
QSG		8-QUINOLINOL SILICA GEL	MATER
QSRR		QUANTITATIVE STRUCTURE RETENTION RELATIONSHIP	METH
QSWB	QSAR	QUANTITATIVE STRUKTUR WIRKUNGSBEZIEHUNG	THEOR CLINCHEM
QUIBEC		BENZYLQUINIDINIUM CHLORIDE	COMP

Abkürzung Akromym abbreviation akronym	Alternative alternative	Bedeutung meaning	Zuordnung related section
R-RNA	r-RNA	RIBOSOMALE RNA	CLINCHEM
RA	RA-MATER	RADIOACTIVE LABELED	COMP
RA		RADIATIVE AUGER	INSTR
RABRA		RUBBER AND PLASTICS RESEARCH ASSOCIATION (GB)	ORGANIS
RAC		RECOMBINANT DNA ADVISORY COMITTEE (NIH,US)	INSTI
RAD	rad	RADIATION ABSORPTION DOSE	METH
RAIRS		RAMAN-INFRARED SPECTROSCOPY	METH
RAM		RANDOM ACCESS MEMORY	ELECT
RAMA		RAMAN MICROANLYSIS	METH
RAMP		RAMAN MICROPROBE	METH
RAP		RUBIDIUM ACID PHTHALATE	COMP
RARE		RESCAUX ASSOCIES POUR LA RECHERCHE EUROPEENE	ORGANIS
RAS		REFLECTION ABSORPTION SPECTROMETRY	METH
RBE	RBW	RELATIVE BIOLOGICAL EFFECTIVENESS	CLINCHEM
RBL	RBL-1	RAT BASOPHIL LEUKEMIA CELLS	CLINCHEM
RBP		RETINOL BINDING PROTEIN	CLINCHEM
RBS	RIBS	RUTHERFORD BACKSCATTERING	METH
RBTC		RHODAMINE B ISOTHIOCYANATE	COMP
RBW	RBE	RELATIVE BIOLOGISCHE WIRKSAMKEIT	CLINCHEM
RCF		RELATIVE CENTRIFUGATIVE FORCE	EVAL
RCLIF	RC-LIF	ROTATIONALLY COOLED LASER INDUCED FLUORESCENCE	METH
RCRA		RESOURCE CONSERVATION AND RECOVERY ACT (US)	REG
RCS		ROYAL CHEMICAL SOCIETY (GB)	ORGANIS
RCSS		RADIAL COMPRESSION SEPARATION SYSTEM	INSTR
RD	R&D ,F&E	RESEARCH AND DEVELOPMENT	
RDA		RECOMMENDED DAILY ALLOWANCES (NIH,US)	REG
RDA		RETRO DIELS-ALDER	METH
RDB		SODIUM DIHYDROBIS-(2-METHOXYETHOXY)-ALUMINATE	COMP
RDE		ROTATED DISC ELECTRODE	INSTR
RDNA	r-DNA	RECOMBINANT DESOXYRIBONUCLEIC ACID	COMP
RDNB		RHODAMINE B NITROBENZENE	COMP
RDS		RELATIVE DETECTION SENSITIVITY	EVAL METH

Abkürzung Akromym abbreviation akronym	Alternative alternative	Bedeutung meaning	Zuordnung related section
RDWBM		RECONSTITUTED DRIED WHOLE BOVINE MILK	CLINCHEM
RE		RESTRICTION ENZYME	CLINCHEM
RE		REFERENCE ELECTRODE	INSTR
REA		RADIOENZYME ASSAY	CLINCHEM
REC		RADIOACTIVE ELECTRON CAPTURE	INSTR
RED		REFLECTION ELECTRON DIFFRACTION	METH
REE		RARE EARTH ELEMENTS	COMP
REELS	HRELS	REFLECTION ELECTRON ENERGY LOSS SPECTROSCOPY	METH
REM		RASTER ELEKTRONEN MIKROSKOPIE	METH
REM	rem	ROENTGEN EQUIVALENT MAN	CLINCHEM REG
REM		REFELCTION ELECTRON MICROSCOPY	METH
REMEDIE		REFLECTION ELECTRON MICROSCOPY AND ELECTRON DIFFRATION AT INTERMEDIATE ENERGIES	METH
REMPI		RESONANTLY ENHANCHED MULTIPHOTON IONIZATION	INSTR
REP		ROENTGEN EQUIVALENT PYSICAL	EVAL
RER		ROUGH ENDOPLASMIC RETICULUM	CLINCHEM
RES		RELATIVE EMISSION SENSITY	METH
REV		REVERSE PHASE EVAPORATION VESICLE	CLINCHEM
RF	Rf-WERT	RATIO OF FRONTS	EVAL
RF		RADIO FREQUENCY	INSTR
RF		RESPONSE FACTOR	METH
RFA		X-RAY FLUORESCENE ANALYSIS	METH
RFI		RADIOFREQENCY INTERFERENCE	INSTR
RFLP		RESTRICTION FRAGMENT LENGTH POLYMORPHISM	CLINCHEM
RFP		RIFAMPICIN	CLINCHEM COMP
RGA		RESIDUAL GAS ANALYSIS	METH
RGC		RECONSTRUCTED GAS CHROMATOGRAMM	EVAL
RGC		RADIO GASCHROMATOGRAPHY	METH
RH	ThTH	THYREOTROPES HORMONE	CLINCHEM
RH		RELATIVE HUMIDITY	EVAL
RHEED		REFLECTION HIGH ENERGY ELECTRON DIFFRACTION	METH
RHF		ROOTHAM -OR RESTRICTED HARTREE FOCK	THEOR
RI		RETENTION INDEX	METH

Abkürzung Akromym	Alternative	Bedeutung	Zuordnung
abbreviation akronym	alternative	meaning	related section
RIA		RADIO IMMUNO ASSAY	METH
RIAC		RAMAN AND INFRARED ANALYTICAL COMMITTEE (JAP)	ORGANIS
RIBE		REACTIVE ION BEAM ETCHING	TECHN
RIBS		RUTHERFORD ION BACKSCATTERING SPECTROSCOPY	METH
RID	R.ID.	REGLEM.INTERN. CONCERNANT LE TRANSPORT DES MERCHANDIES DANGEREUREUSES	REG
RIE		REACTIVE ION ETCHING	TECHN
RIGG	RIgG	RABBIT IMMUNOGLOBULINE	CLINCHEM
RIM		REACTION INJECTION MOULDING PROCEDURE	TECHN POLYM
RIM		REACTANT ION MONITORING	METH
RIM		RASTER IONEN MIKROSKOPIE	METH
RIM		RADIO INDUCTION METHOD	METH
RIM		RETENTION INDEX METHOD	EVAL
RIMS		RESONANCE IONIZATION MASS SPECTROSCOPY	METH
RIS		RESONANCE IONIZATION SPECTROSCOPY	METH
RISKA		RISIKEN IN CHEMISCHEN ANLAGEN (INFUCHS,D)	LIT TECHN
RIU	RIU/FS	REFRACTIVE INDEX / FULL SCALE	INSTR
RIWA		RIJNCOMMISSIE WATERLEIDINGSBEDRIJVEN (NL)	INSTI
RIZA		RIJSINSTITUUT VOOR ZNIVERING VON AFALWATEN (NL)	INSTI
RKM		ROENTGEN KONTRAST MITTEL	CLINCHEM
RL		RADIO LUMINESCENCE	METH
RLCC		ROTATION LOCULAR COUNTER CURRENT CHROMATOGRAPHY	METH
RM		REFERENCE MATERIAL	MATER
RMAED		ROCKING MICRO-AREA ELECTRON DIFFRACTION	METH
RMEE		ROTATING MERCURY FILM ELECTRODE	INSTR
RMM		REMAZOL BRILLIANT BLUE(-DYED)	CLINCHEM COMP
RMS		ROOT MEAN SQUARE	EVAL
RMS		ROYAL MICROSCOPY SOCIETY (GB)	ORGANIS
RMZ	R-M-Z	REICHELT -MEISSL ZAHL	THEOR
RNA		RIBONUCLEIC ACID	CLINCHEM

Abkürzung Akromym abbreviation akronym	Alternative alternative	Bedeutung meaning	Zuordnung related section
RNAA		RADIOACTIVE NEUTRON ACTIVATION ANALYSIS	METH
RNASE		RIBONUCLEIC ACID LEASE	CLINCHEM
ROA		RAMAN OPTICAL ACTIVITY	EVAL
RNP		RIBONUCLEOPROTEIN	CLINCHEM
RNT		RADIO NUCLID TECHNIC	METH
RO	UO	REVERSE OSMOSIS	METH
ROA		RAMAN OPTICAL ACTIVITY	METH
ROC		RESIDUAL ORGANIC CARBON	COMP
RODAC		REPLICATE ORGANISM DIRECT CONTACT TECHNIQUE	CLINCHEM
ROM		READ ONLY MEMORY	ELECT
ROZ		RESEARCH OCTAN ZAHL	TECHN
RP		RESOLVING POWER	INSTR
RPA		RETARDING POTENTIAL ANALYSER	INSTR
RPC		REVERSE(D) PHASE CHROMATOGRAPHY	METH
RPED		ROCKING PROBE ELECTRON DIFFRACTION	METH
RPFC		REVERSED FLASH CHROMATOGRAPHY	METH
RPIPP	RP-IPP	REVERSE(D)PHASE ION PAIR CHROMATOGRAPHY	METH
RPL		RADIOPHOTOLUMINESCENCE	METH
RPM	rpm	ROUNDS (REVOLUTIONS) PER MINUTE	INSTR
RPM		REVERSED PHASE MATERIAL	MATER
RPN	UPN	REVERSE POLISH NOTATION	ELECT
RPS		RANDOM PULSE SEQUENCE	INSTR
RR		RELATIVE REPRESSION	METH
RRF		RELATIVE RESPONSE FACTOR	EVAL METH
RRKM		RICE,RAMSBERGER,KASSEL UND MARKUS	THEOR
RRS		RESONANCE RAMAN SCATTERING SPECTROMETRY	METH
RRSB		DISTRIBUTION TO ROSIN, RAMMLER, SPERLING AND BENETT	THEOR
RRT		RELATIVE RETENTION TIME	EVAL
RSA	RSD	RELATIVE STANDARD ABWEICHUNG	EVAL
RSAED		ROCKING SELECTED AREA ELECTRON DIFFRACTION PATTERN	EVAL
RSC		ROYAL SOCIETY OF CHEMISTRY (GB)	ORGANIS
RSD	RSA	RELATIVE STANDARD DEVIATION	EVAL
RSF		RELATIVE SENSITIVITY FACTOR	INSTR

Abkürzung Akromym	Alternative	Bedeutung	Zuordnung
abbreviation akronym	alternative	meaning	related section
RSK		REAKTOR SICHERHEITS KOMMISSION (BMI,D)	INSTI
RSK		RICHTWERTE UND SCHWANKUNGSBREITE BESTIMMTER KENNZAHLEN	EVAL
RSMW		RAPID SCANNING MULTIPLE WAVELENGTH	INSTR
RSPA	RoSPA	ROYAL SOCIETY FOR THE PREVENTION OD ACCIDENTS (GB)	INSTI
RSV		RESPIRATORY SYNCYTIAL VIRUS	CLINCHEM
RT		REVERSE TRANSCRIPTION	CLINCHEM
RTD	WLD	RESITANCE TEMPERATURE DETECTOR	INSTR
RTECS		REGISTRY OF TOXIC EFFECTS OF CHEMICAL SUBSTANCES (US)	REG
RTL		RESISTOR TRANSISTOR LOGIC	ELECT
RTM		RASTER-TUNNEL MICROSCOPY	METH
RTM		REAL TIME MONITORING	INSTR
RTP		ROOM TEMPERATURE PHOSPHORESCENE	METH
RTPL		ROOM TEMPERATURE PHOSPHORESCENCE IN LIQUID CHROMATOGRAPHY	METH
RTV		ROOM TEMPERATURE VULCANIZATION	TECHN
RWDPP		RADIOACTIVE WASTE DISPOSAL PILOT PLANT	TECHN
RWTÜV		RHEINISCH-WESTFÄLISCHER TECHNISCHER ÜBERWACHUNGSVEREIN eV (ESSEN,D)	COMPANY

Abkürzung Akromym abbreviation akronym	Alternative alternative	Bedeutung meaning	Zuordnung related section
SA		SURFACE ANALYSIS	METH
SAA		SUCCINIC ANHYDRIDE	COMP
SAA		STANDARDS ASSOCIATION OF AUSTRALIA (AUS)	INSTI
SAAC		SUCCINILIERTE AMINOALKYL CELLULOSE	MATER
SAC	SAC-435	SPECIFIC ABSORPTION COEFFIZIENT - WAVELENGTH	EVAL
SACI		SOCIETY OF AGRICULTURAL CHEMICAL INDUSTRY (JAP)	ORGANIS
SAD		SINGLE ATOM DETECTION	METH
SAD		SELECTIVE AREA DIFFRACTION	METH
SADP		N-SUCCININIDYL-(4-AZIDOPHENYLDITHIO)-PROPIONATE	COMP
SAECP		SELECTED AREA ELECTRON CHANNELING PATTERNS	EVAL
SAED		SELECTIVE AREA ELECTRON DIFFRACTION	METH
SAEP		SENIOR ADVISERS TO ECE GOVERNMENTS ON ENVIRONMENTAL PROBLEMS	INSTI
SAES		SCANNING AUGER ELECTRON SPECTROSCOPY	METH
SAGO		SOCIETA DI RICERCA PER L'ORGANIZZAZIONE SANITARIA (I)	ORGANIS
SAGUF		SCHWEIZERISCHE ARBEITSGEMEINSCHAFT UMWELTFORSCHUNG (CH)	ORGANIS
SAH		S-ADENOSYL-L-HOMOCYSTEIN	COMP
SAIB		SUCROSE ACETATE ISOBUTYRATE	MATER
SAICAR	suc-AICAR	5-AMINO-4-IMIDAZOLE CARBOXAMIDE RIBONUCLEOTIDE	COMP
SAIDS		SIMIAN ACQUIRED IMMUNO DEFICIENCY SYNDROME	CLINCHEM
SAL		SALSALINOL-HYDROBROMIDE	COMP
SAM		SCANNING AUGER MICROPROBE	METH
SAM		SCANNING ACOUSTIC MICROSCOPY	METH
SAM		S-ADENOSYL-L-METHIONINE	COMP
SAME	SAMe	S-ADENOSYL-1-METHIONINE	CLINCHEM COMP
SAMW		SCHWEIZERISCHE AKADEMIE DER MEDIZINISCHEN WISSENSCHAFTEN (CH)	ORGANIS
SAN		STYRENE-ACRYL-NITRILE	MATER POLYM
SANS		SMALL ANGLE NEUTRON SCATTERING	METH

Abkürzung Akromym abbreviation akronym	Alternative alternative	Bedeutung meaning	Zuordnung related section
SANZ		SCHWEIZERISCHE ARZNEIMITTELNEBENWIRKUNGSZENTRALE (CH)	INSTI
SAPO		SILICIUM-ALUMINIUM-PHOSPHATE	MATER
SAQ		SCHWEIZER. ARBEITSGEMEINSCHAFT QUALITÄTSFÖRDERUNG (CH)	ORGANIS
SAR		SPECIFIC ABSORPTION RATE	EVAL MATER
SAS		SOCIETY FOR APPLIED SPECTROSCOPY (US)	ORGANIS
SAS		SURFACE ACTIVE SUBSTANCES	COMP
SASP		SCHWEIZERISCHE ARBEITSGEMEINSCHAFT SPECTRAL ANALYSE (CH)	ORGANIS
SATO		SELF ALIGNED THICK OXID	MATER ELECT
SATW		SCHWEIZERISCHE AKADEMIE DER TECHNISCHEN WISSENSCHAFTEN (CH)	INSTI
SAW		SURFACE ACOUSTIC WAVE	EVAL
SAX		STRONG ANION EXCHANGER	MATER
SAX		SELECTED AREAS X-RAY PHOTOELECTRON SPECTROSCOPY	METH
SAXS		SMALL ANGLE X-RAY SCATTERING	METH
SBA		SOYABEAN-AGGLUTHININE	CLINCHEM
SBDF	SBD-F	7-FLUORO-2-OXA-1,3-DIAZOLE-4-SULFONIUM	COMP
SBH		SODIUM BOROHYDRIDE	COMP
SBL		STATENS BAKTERIOLOGISKA LABORATORIUM STOCKHOLM (S)	INSTI
SBPI		SOCIETY OF BRITISCH PRINTING INK. (GB)	ORGANIS
SBR	SRV	SIGNAL TO BACKGROUND RATIO	EVAL
SBR		STYRENE-BUTADIENE RUBBER	MATER
SBV		SÄURE BINDUNGS VERMÖGEN	EVAL
SC		SÄULEN CHROMATOGRAPHIE	METH
SC		SUPER CAPACITANCE	INSTR
SCAD		SURFACE CHARACTERIZATION AND DEPTH POFILING	METH
SCANIIR		SURFACE COMPOSITION BY ANALYSIS OF NEUTRAL AND ION IMPACT RADIATION	METH
SCAO		SPERICAL CLOUD ATOMIC ORBITALS	THEOR
SCC		GESELLSCHAFT SCHWEIZERISCHER COSMETIK CHEMIKER (CH)	ORGANIS
SCCB		SCHWEIZER.COORDINATIONS COMMITTEE FÜR BIOTECHNOLOGIE (CH)	INSTI

Abkürzung Akromym	Alternative	Bedeutung	Zuordnung
abbreviation akronym	alternative	meaning	related section
SCDF		SCANNING ELECTRON DIFFRACTION	METH
SCE	GKE	SATURATED CALOMEL ELECTRODE	INSTR
SCE		SOMATIC CELLS EFFECT	CLINCHEM
SCE		SISTER CHROMATIDINE EXCHANGE	CLINCHEM
SCE		SECONDARY CHEMICAL EQUILIBRIUM	THEOR
SCEM		SINGLE-CHANNEL ELECTRON MULTIPLIER	INSTR
SCEP		SELF CONSISTENT ELECTRON PAIRS	THEOR
SCF		SELF CONSISTENT FIELD	THEOR
SCFM	scfm	STANDARD CUBIC FEET PER MINUTE	EVAL
SCG		SUPER CRITICAL GAS	MATER
SCG		SCHWEIZERISCHE CHEMISCHE GESELLSCHAFT (CH)	ORGANIS
SCHV	SChV	SCHWEIZERISCHER CHEMIKER-VERBAND (CH)	ORGANIS
SCI		SOCIETY OF CHEMICAL INDUSTRY (INT EUR)	ORGANIS
SCIC		SINGLE COLUMN ION CHROMATOGRAPHY	METH
SCIM	STEM	SCANNED TRANSMISSION ELECTRON IMAGE MICROSCOPY	METH
SCLC		SEQUENCE CENTRIFUGAL LAYER CHROMATOGRAPHY	METH
SCLO		SELF CONSISTENT LINEAR ORBITAL	THEOR
SCO		STUDENTEN-COMMITTEE FÜR UMWELT ÖKONOMIK (CH)	ORGANIS
SCOPE		SCIENTIFIC COMMISION ON PROBLEM OF THE ENVIRONMENT (US)	ORGANIS
SCOR		SCIENTIFIC COMMITTEE ON OCEANIC RESEARCH (ICSU)	ORGANIS
SCOT		SURFACE COATED OPEN TUBULAR COLUMN	MATER
SCP		SINGLE CELL PROTEIN	CLINCHEM
SCP		SODIUM CELLULOSE PHOSPHATE	COMP
SCPS		SULFOCHLORO PHENOL S	COMP
SCR		SELECTIVE CATALYTIC REDUCTION	COMP TECHN
SCRS		STOKES COHERENT RAMAN SPECTROSCOPY	METH
SCT		SURFACE CHANGE TRANSISTOR	ELECT
SCX		STRONG CATION EXCHANGER	MATER
SDA		SPERICAL DEFLECTOR ANALYSER	INSTR
SDA		SOAP AND DETERGENT ASSOCIATION (US)	ORGANIS
SDAC		SOAP AND DETERGENT ASSOCIATION OF CANADA (CDN)	ORGANIS

Abkürzung Akromym abbreviation akronym	Alternative alternative	Bedeutung meaning	Zuordnung related section
SDC		SOCIETY OF DYERS AND COLORISTS (GB)	ORGANIS
SDE		SIMULTANOUS (STEAM)DESTILLATION EXTRACTION	METH
SDIM		SYSTEM DOKUMENTATION UND INFORMATION METALLURGIE (D)	LIT
SDLTS		SCANNING DEEP LEVEL TRANSIENT SPECTROSCOPY	METH
SDMH		SYMMETRIC DIMETHYL HYDRAZINE	COMP
SDMM		SCANNING DESORPTION MOLECULE MICROSCOPY	METH
SDP		4,4'-SULFONYLDIPHENOL	COMP
SDPP		N-SUCCINIMIDYL-DIPHENYL PHOSPHATE	COMP
SDS		SODIUM DODECYL SULFATE	COMP
SDVB	S-DVB	STYRENE DIVINYL BENZENE COPOLYMER	MATER
SE		SIZE EXCLUSION	METH
SE		STAPHYLOKOKKEN-ENDOTOXINE	CLINCHEM
SE		SECONDARY ELECTRON / EMISSION	METH
SEA		SOUND EMISSION ANALYSIS	METH
SEA		SPEZIFIC ELISA ACTIVITY	CLINCHEM
SEB		SECONDARY ELECTRON BREMSSTRAHLUNG	METH
SEC		SIZE EXCLUSION CHROMATOGRAPHY	METH
SEC		SPECTRO ELECTROCHEMISTRY	METH
SECOTOX		SOCIETY OF ECOTOXICOLOGY AND ENVIRONMENTAL SAFETY (INT)	ORGANIS
SECS		SELECTIVE ELECTRON CAPTURE SENSITIZATION	METH
SED		SECONDARY ELECTRON DETECTOR	INSTR
SEE		SECONDARY ELECTRON EMISSION	METH
SEE		SOCIETY OF ENVIRONMENTAL ENGINEERS (GB)	ORGANIS
SEFEL		SECRET.EUROP.POUR LA FABRICATION DES EMBALLAGES METALLIQUES LEGERS	INSTI
SEFT		SPIN ECHO FOURIER TRANSFORM NMR	METH
SEG		SIDE ENTRY GONIOMETER	INSTR
SEL		SOUND EXPOSURE LEVEL	TECHN
SEM	REM	SCANNING ELECTRON MICROSCOPY	METH
SEM	SEV	SECONDARY ELECTRON MULTIPLIER	INSTR
SEM		SOCIETY FOR EXPERIMENTAL MECHANICS (US)	ORGANIS

Abkürzung Akromym abbreviation akronym	Alternative alternative	Bedeutung meaning	Zuordnung related section
SEM-EBIC		SCANNING ELECTRON MICROSCOPY-ELECTRON BEAM CURRENT	METH
SEM-EDS		SCAN. ELECTR. MICROS - ENERGY DIPERSIVE X-RAY SPECTROSCOPY	METH
SEMCL	SEM-Cl	CHLOROMETHYL TRIMETHYLSILYL ETHYLESTER	COMP
SEPC		SERVICE DE L'ENVIRONNEMENT ET DE PROTECTION DES CONSOMMATEURS (EG)	INSTI
SEPES		SYNCHROTRONRADIATION EXCITED PHOTOELECTRON SPECTROSCOPY	METH
SEPIL		SELECTIVELY EXCITING PROBE ION LUMINESCENCE	METH
SER		SURFACE ELEVATED RAMAN EFFECT	INSTR
SER	Ser	SERINE	COMP
SER		SMOOTH ENDOPLASMATIC RETICULUM	CLINCHEM
SERC		SCIENCE AND ENGINEERING RESEARCH COUNCIL (GB)	INSTI
SERRS		SURFACE ENHANCED RESONANCE RAMAN SCATTERING	METH
SERS		SURFACE ENHANCHED RAMAN SPECTROSCOPY	METH
SES	2,4-DES	DISUL SODIUM	COMP
SESD		SCANNING ELECTRON STIMULATED DESORPTION	METH
SET		SINGLE EXPOSURE TECHNIQUE	METH
SET		STAPHYLOKOKKEN-ENTEROTOXIN	CLINCHEM
SETAC		SOCIETY OF ENVIRONMENTAL TOXICOLOGY AND CHEMISTRY (US)	ORGANIS
SEV	SEM	SEKUNDAERELEKTRONEN VERVIELFACHER	INSTR
SEV		SCHWEIZERISCHER ELEKTROTECHNISCHER VEREIN (CH)	ORGANIS
SEX		SODIUM ETHYL XANTHATE	COMP
SEXAFS		SURFACE EXTENTED X-RAY ABSORPTION FINE STRUCTURE	METH
SF		SPONTANFISSION = SPONTANSPALTUNG	METH
SFB		SONDER FORSCHUNGS BEREICH (DFG/D)	INSTI
SFBC		SOCIETE FRANCAISE DE BIOLOGIE CHEMIQUE (F)	ORGANIS
SFC		SUPERCRITICAL FLUID CHROMATOGRAPHY	METH
SFC	S.F.C	SOCIETE FRANCAISE DE CHEMIE (F)	ORGANIS

Abkürzung Akromym	Alternative	Bedeutung	Zuordnung
abbreviation akronym	alternative	meaning	related section
SFEP		SELFCONSISTENT ELECTRON PAIRS	THEOR
SFFF		SEDIMENTATION FIELD FLOW FRACTIONATION	METH
SFT		STATENS FORURENSNINGSTILSYN (N)	INSTI
SG		SECTOR GROUP (CEFIC)	ORGANIS
SGA	ASSPA	SCHWEIZERISCHE GESELLSCHAFT FUER AUTOMATION (CH)	ORGANIS
SGA		SINGLE GRID ALIGMENT	ELECT TECHN
SGAAC		SCHWEIZ.GESELLSCHAFT F. ANALTYISCHE UND ANGEWANDTE CHEMIE (CH)	ORGANIS
SGB	SSB	SCHWEIZERISCHE GESELLSCHAFT FUER BIOCHEMIE (CH)	ORGANIS
SGCI	SSIC	SCHWEIZER GESELLSCHAFT FÜR CHEMISCHE INDUSTRIE (CH)	ORGANIS
SGF		SMALL GENE FRAGMENTS	CLINCHEM
SGKC	SSCC	SCHWEIZERISCHE GESELLSCHAFT FUER KLINISCHE CHEMIE (CH)	ORGANIS
SGIM		SCHWEIZ. GESELLS. F. INSTRUMENTALANALYTIK UND MIKROCHEMIE (CH)	ORGANIS
SGLH		SCHWEIZ. GESELLS. F. LEBENSMITTELHYGIENE (CH)	ORGANIS
SGM	SSM	SCHWEIZERISCHE GESELLSCHAFT FUER MIKROBIOLOGIE (CH)	ORGANIS
SGOS		SILICON GATE OXIDE SEMICONDUCTOR	ELECT
SGRPA		SPHERICAL GRID RETARDING POTENTIAL ANALYSER	METH
SGT		SCHWEIZER GALVANOTECHNISCHE GESELLSCHAFT (CH)	ORGANIS
SGTM		SECTOR GROUP TITAIUM/DIOXIDE MANUFACTURERS (CEFIC)	ORGANIS
SGU		SCHWEIZERISCHE GESELLSCHAFT FÜR UMWELTSCHUTZ (CH)	ORGANIS
SGZP		SCHWEIZER GESELLSCHAFT ZERSTÖRUNGSFREIE PRÜFUNG (CH)	ORGANIS
SHBG		SEX HORMONE BINDING GLOBULINE	CLINCHEM
SHEED		SCANNING (OR SECONDARY) HIGH ENERGY ELECTRON DIFFRACTION	METH
SHF		SUPER HIGH FREQUENCY	INSTR
SHG		SECOND HARMONIC GENERATION	THEOR

Abkürzung Akromym abbreviation akronym	Alternative alternative	Bedeutung meaning	Zuordnung related section
SHIP		SEPARATOR FOR HEAVY ION REACTION PRODUCTS	INSTR TECHN
SHM		SIMPLE HARMONIC MOTION	THEOR
SHOP		SHELL HIGHER OLEFIN PROCESS	TECHN
SI		SYSTEME INTERNATIONAL DES UNITS	NORM
SI		SURFACE IONIZATION	METH
SIA		SCHWEIZERISCHER INGENIEUR- UND ARCHITEKTEN VEREIN (CH)	ORGANIS
SIAO		SCHWEIZER. INTERESSENGEMEINS. DER ABFALLBESEITIGUNGSORG. (CH)	ORGANIS
SIC		STANDARD INDUSTRIAL CLASSIFICATION (US)	TECHN
SICP		SELECTED ION CURRENT PROFILE	EVAL
SID		SINGLE / SELECTED ION DETECTION	METH
SID		SURFACE INDUCED DISSOCIATION	METH
SIEM		SHADOW IMAGE ELECTRON MICROSCOPY	METH
SIIMS		SECONDARY ION IMAGING MASS SPECTROSCOPY	METH
SIM		SELECTED ION MONITORING	INSTR
SIMA		THE SCIENTIFIC INSTRUMENT MANUFACTURERS ASSOCIATION	ORGANIS
SIMA		SECONDARY ION MICROANALYSIS	METH
SIMMS	IMMA	SECONDARY ION MICROPROBE	METH
SIMS		SECONDARY ION MASSSPECTROMETRY	METH
SIMS-IDP		SECONDARY IONS MASSSPECTROSCOPY IMAGE DEPTH PROFILING	EVAL
SIN		SIMULTANEOUS INTERPENETRATING NETWORK	MATER POLYM
SIN		SCHWEIZERISCHES INSTITUT FÜR NUKLEARFORSCHUNG (CH)	INSTI
SIPOS		SEMI-INSULATED POLYCRYSTALINE SILICON	ELECT MATER
SIPS		SPUTTER INDUCED PHOTON SPECTROSCOPY	METH
SIR		SELECTED ION RECORDING	EVAL
SIRMCE		SOCIETY INTERN.FOR RESEARCH ON CIVILIZATION DIEASES AND ENVIRONMENT	ORGANIS
SIRRS		SURFACE INDUCED RESONANT SCATTERING	METH
SISO		SINGLE INPUT-SINGLE OUTPUT	TECHN
SIT		SELF INDUCED TRANSPARENCY	METH

abbreviation akronym	alternative	meaning	related section
SITS		4-ACETAMIDO-4'-ISOTHIOCYANATO-STILBENE-2,2'-DISULFONIC ACID	COMP
SIX		STRONG ION EXCHANGER	MATER
SKB		SCHWEIZ. KOORDINATIONSAUSSCHUß F BIOTECHNOLOGIE (CH)	ORGANIS
SKE		STEINKOHLE EINHEITEN	TECHN
SKI		SVERIGES KEMISKA INDUSTRIEKONTOR STOCKHOLM (S)	INSTI
SKI		SCHWEIZERISCHES KRANKENISTITUT (CH)	INSTI
SLAC		STRATFORD LINEAR ACCELERATOR CENTER	INSTI
SLAM		SCANNING LASER ACOUSTIC MICROSCOPY	METH
SLEEP		SCANNING LOW ENGERGY ELECTRON PROBE	METH
SLGC	SL-GC	SPLITLESS GAS CHROMATOGAPHIC TECHNIC	METH
SLM		SOUND LEVEL METER	INSTR
SLR		SUPER LATTICE REFLECTION	METH
SLS		STEARO-LINOLEO STEARINE	MATER
SLS		SODIUM LAURYL SULFATE	COMP
SLT		SECOND LAW OF THERMODYNAMICS	THEOR
SM		STPREPTOMYCIN	CLINCHEM
SMA		SPHERICAL MIRROR ANALYZER	INSTR
SMA		SIMULTENOUS MULTIWAVELENTH ACCQUISITION	INSTR
SMAD		SOLVATED METAL ATOM DISPERSION	COMP
SMC		SHEET MOULDING COMPOUND	MATER POLYM
SMCC		N-SUCCINIMIDYL-4(MALEIMIDOMETHYL CYCLOHEXANE)-1-CARBOXYLATE	COMP
SMDE		STATIONARY MERCURY DROP ELECTRODE	INSTR
SMEAH		SODIUM BIS(2-METHOXYETHOXY)ALUMINIUMHYDRIDE	COMP REAG
SMGE		SELECTIVE MULTISOLVENT GRADIENT ELUTION	METH
SMIM		SELECTIVE METASTABLE ION MONITORING	EVAL
SMON	S.M.O.N.	SUBACUTE MYELO-OPTIC NEUROPATHY SYNDROME	CLINCHEM
SMP		S(-)-2-(METHOXYMETHYL)-PYRROLIDINE	COMP
SMPB		SUCCINIMIDYL-4-(p-MALEIMIDOPHENYL)-BUTYRATE	COMP
SMS		STYRENE-METHYLSTYRENE-COPOLYMER	MATER POLYM
SN		SCHWEIZER NORM (CH)	NORM

Abkürzung Akromym abbreviation akronym	Alternative alternative	Bedeutung meaning	Zuordnung related section
SN	s/n	SIGNAL TO NOISE RATIO	METH
SNADANS		2-(4-SULFO-NAPHTHYLAZO-)-1,8-DIHYDROXY NAPHTHALIN-3,6-DISULFONIC ACID	COMP
SNG		SCHWEIZERISCHE NATURFORSCHENDE GESELLSCHAFT (CH)	ORGANIS
SNG		SYNTHETIC NATURAL GAS	MATER TECHN
SNIC		SOCIETE NATIONALE DES INDUSTRIES CHIMIQUES (ALG)	ORGANIS
SNMS		SPUTTERED NEUTRAL MASS SPECTROSCOPY	METH
SNMS		SECONDARY NEUTRALIZATION MASS SPECTROSCOPY	METH
SNPA		N-SUCCININIDYL-p-NITRO-PHENYL ACETATE	COMP
SNQ		SPALLATIONS NEUTRONEN QUELLE (NOT REALIZED KFA JUELICH,D)	TECHN
SNR	s/n SRV	SIGNAL TO NOISE RATIO	EVAL
SNV		SCHWEIZERISCHE NORMEN VEREINIGUNG (CH)	ORGANIS
SO		SPIN ORBITAL	THEOR
SOCMA		SYNTHETIC ORGANIC CHEMICAL MANUFACTURERS ASSOCTIATION (US)/	ORGANIS
SOD		SUPEROXID DISMUTASE	CLINCHEM
SOEH		SOCIETY OF OCCUPATIONAL AND ENVIRONMENTAL HEALTH	ORGANIS
SOLZ		SECOND ORDER LAUE ZONE	EVAL
SOMO		SINGLE OCCUPIED MOLECULAR ORBITAL	THEOR
SONAR		SOUND NAVIGATING AND RANGING	TECHN
SONRES		SATURATED OPTICL NONRESONANT EMISSION SPECTROSCOPY	METH
SOS		STEARO-OLEO STEARINE	MATER
SOS		SILICON ON SPAPHIRE	ELECT MATER
SOT		SOCIETY OF TOXICOLOGY (US)	ORGANIS
SP		SPECTRA PHYSICS	COMPANY
SPA		SUPER PHOSPHORIC ACID	COMP
SPA		STAPHYLOCOCCUS AUREUS PROTEIN A	CLINCHEM
SPADNS		2-(p-SULFOPHENYLAZO)1,8-DIHYDROXY-3,6-NAPHTHALENE DISULFONIC ACID	COMP
SPC		SELF POLISHING ANTIFAULING COLOURS	MATER
SPCA		SERUM PROTHROMBIN CONVERSION ACCELERATOR	CLINCHEM
SPCD		SYNCHRONANSLY PUMPED CAVITY DUPED	INSTR

Abkürzung Akromym	Alternative	Bedeutung	Zuordnung
abbreviation akronym	alternative	meaning	related section
SPD		SCALED PARTICLE THEORY	THEOR
SPDP		N-SUCCINIMIDYL-3-(2-PYRIDYLDITHIO)-PROPIONATE	COMP
SPE		SOLID PHASE EXTRACTION	METH
SPE		SERUM PROTEIN ELECTROPHORESIS	METH
SPEAR		STANFORD POSITRON ELECTRONACCELERATOR RING (US)	INSTI
SPEG		SOLID PHASE EPITAXIAL GROWTH	MATER
SPF		SPECIFIED PATHOGEN FREE	CLINCHEM
SPF		SVERIGES PLASTFÖRBUND (S)	ORGANIS
SPG		SYNCHRONOUS PULSE GENERATOR	INSTR
SPI		SELECTIVE POPULATION INVERSION NMR	METH
SPI		SOCIETY OF PLASTICS INDUSTRY (US)	ORGANIS
SPIET		SOCIETE DE PROMOTION DE L'IND.DE L'ENNOBLISSEMENT TEXTILE PLASTIQUES	ORGANIS
SPIXE		SCANNING PROTON INDUCED X-RAY EMISSION	METH
SPMP		SYNDICAT PROFESSIONEL DES PRODUCTEURS DE MATIERS	ORGANIS
SPP		SPIN POLARIZED PHOTO EMISSION	METH
SPPS		SOLID PHASE PEPTIDE SYNTHESIS	METH
SPR		SOLID PHASE REACTOR	INSTR TECHN
SPRENGG	SprengG	GESETZ ÜBER EXPLOSIONSGEFAEHRLICHE STOFFE (D)	REG
SPT		SELECTIVE POPULATION TRANSFER NMR	METH
SPT		BRANCHEFORENING FOR SAEBE-PARFUMERI,TOILET-OG KEMISK TEKNISKE ARTIKLER (DK)	ORGANIS
SQDG		SULFOQUINO-DANSYL GLYCERIDE	COMP
SQS		SCHWEIZER VEREINIGUNG QUALITÄTSSICHERUNG (CH)	ORGANIS
SQUID		SUPERCONDUCTING QUANTUM INTERFERENCE DETECTOR	INSTR
SR		SYNCHROTRON RADIATION	METH
SRC		SCIENCE RESEARCH COUNCIL (GB)	INSTI
SRC		SOLVENT REFINED COAL	MATER
SRET		SCANNING REFERENCE ELECTROCHEMICAL TECHNIQUE	METH
SRI		STANFORD RESEARCH INSTITUTE (GB)	INSTI
SRM		STANDARD REFERENCE MATERIAL	MATER

Abkürzung Akromym	Alternative	Bedeutung	Zuordnung
abbreviation akronym	alternative	meaning	related section
SRRT		SCHWEIZERISCHE GESELLSCHAFT FÜR REINRAUMTECHNIK (CH)	ORGANIS
SRS		SLOW REACTING SUBSTANCES	COMP
SRS		STIMULATED RAMAN SPECTROSCOPY	METH
SRS		SURFACE REFLECTANCE SPECTROSCOPY	METH
SRV	SNR	SIGNAL ZU RAUSCH VERHÄLTNIS	METH
SRWS		STANDARD REFERENCE WATER SAMPLE	METH
SRXRF		SYNCHROTRON RADIATION X-RAY FLUORESCENCE	METH
SS	SS-ALLOY	STAINLESS STEEL	MATER
SSAGAR	SS Agar	SALMONELLA SHIGELLA AGAR	CLINCHEM
SSB	SGB	SOCIETE SUISSE DE BIOCHEMIE (CH)	ORGANIS
SSB		SACHVERSTAENDIGENKOMMISSION F.FRAGEN DER SICHERUNG DES KERNBRENNSTOFFKREISLAUFES (D)	INSTI
SSCC	SGKC	SOCIETE SUISSE DE CHEMIE CLINIQUE (CH)	ORGANIS
SSFLC		SURFACE-STABALIZED FERROELECTRIC LIQUID CRYSTALS	MATER
SSI		SMALL-SCALE INTEGRATION	ELECT INSTR
SSI		STATENS SERUMINSTITUT KOPENHAGEN (DK)	INSTI
SSIC	SGCI	SOCIETE SUISSE DES INDUSTRIES CHIMIQUES (CH)	ORGANIS
SSIMS		STATIC SECONDARY ION MASS SPECTROSCOPY	METH
SSK		STRAHLEN SICHERHEITS KOMMISSION (BMI,D)	INSTI
SSLC		SUPER SPEED LIQUID CHROMATOGRAPHY	METH
SSM	SGM	SOCIETE SUISSE DE MICROBIOLOGIE (CH)	ORGANIS
SSMS	FMS	SPARK SOURCE MASS SPECTROMETRY	METH
SSP		1,2-DISTEROYLPALMITIN	COMP
SSP		SUSPENDED SOLID PHASE SEPARATION	METH
SSRL		STANFORD SYNCHROTRON RADIATION LABORATORY (US)	INSTI
SSRT		SLOW STRAIN RATE TEST	MATER METH
SSS		TRISTEARINE	MATER
SSS		SODIUM STYRENE SULFONATE	COMP
SSS		SIEDEHITZE , SEITENKETTE , SUBSTITUTION	METH
SSS		SEQUENTIAL SLOW SCANNING	METH

Abkürzung Akromym	Alternative	Bedeutung	Zuordnung
abbreviation akronym	alternative	meaning	related section
STADA		STANDARDPRÄPARATE DEUTSCHER APOTHEKER (D)	MATER
STAT		SLOTTED TUBE ATOM TRAP	INSTR
STAWA	StAWA	STAATLICHES AMT FÜR WASSER UND ABFALL (NRW,D)	INSTI
STBA	StBA	STATISTISCHES BUNDESAMT (WIESBADEN,D)	INSTI
STEBIC		SCANNING TRANSMISSION ELECTRON BEAM CURRENT	METH
STEL		SHORT TERM EXPOSURE LIMIT	REG
STEM	SCIM	SCANNING TRANSMISSION ELECTRON MICROSCOPY	METH
STESR		SATURATION-TRANSPORT-ELECTRONSPIN RESONANCE	METH
STF		STABILIZED TEMPERATURE FURNACE	INSTR
STH		SOMATROPES HORMON	CLINCHEM
STIC		SCIENTIFIC,TECHNOLOGICAL AND INTERNATIONAL AFFAIRS (NSF,US)	INSTI
STK		SCHWEIZERISCHE GESELLSCHAFT F. THERMOANALYTIK UND KALORIMETRIE	ORGANIS
STO		SLATER TYPEORBITALS	THEOR
STP	DOM	2,5-DIMETHOXY-4-METHYL AMPHETAMINE	CLINHEM COMP
STP		PENTA-SODIUM TRIPHOSPHATE	COMP
STPF		STABILIZED TEMPERATURE PLATFORM FURNACE	INSTR
STPP		SODIUM TRIPOLYPHOSPHATE	COMP
STRSCHV	StrSchV	STRAHLENSCHUTZVERORDNUNG (D)	REG
STRUP		SHROUD OFTURIN RESEARCH PROJECT	TECHN
STTL		SCHOTTKY TRANSISTOR TRANSISTOR LOGIC	ELECT
STTL		SUOMEN TEKSTIILITENILLIDEN LIITON VÄR JÄYS-JA VIEMEISTYSSJAOSTO (FIN)	ORGANIS
SUVA		SCHWEIZERISCHE UNFALLVERSICHERUNGSANSTALT (LUZERN,CH)	INSTI
SV		STRIPPING VOLTAMMETRY	METH
SV		STIFTERVERBAND FÜR DIE DEUTSCHE WISSENSCHAFT (D)	ORGANIS
SVA		SCHWEIZER VEREINIGUNG DER ARBEITSHYGIENIKER (CH)	ORGANIS
SVCC		SCHWEIZERISCHER VEREIN DER CHEMIKER COLORISTEN (CH)	ORGANIS

Abkürzung Akromym	Alternative	Bedeutung	Zuordnung
abbreviation akronym	alternative	meaning	related section
SVCT		SCHWEIZERISCHE VEREINIGUNG DIPLOMIERTER CHEMIKER HTL (CH)	ORGANIS
SVDB		SCHWEIZERISCHER VEREIN FÜR DRUCKBEHÄLTERÜBERWACHUNG	NORM
SVEFF		SVERIGES FÄRGFABRIKANTERS FÖRENING (S)	ORGANIS
SVF		SCHWEIZERISCHE VEREINIGUNG VON FÄRBEREIFACHLEUTEN (CH)	ORGANIS
SVG		SCHWEIZERISCHE VEREINIGUNG FÜR GESUNDHEITSTECHNIK (CH)	ORGANIS
SVGW		SCHWEIZERISCHER VEREIN VON GAS-UND WASSERFACHMÄNNERN (CH)	ORGANIS
SVLFC		SCHWEIZERISCHER VEREINIGUNG DER LACK- UND FARBENCHEMIKER (CH)	ORGANIS
SVMT		SCHWEIZERISCHER VERBAND FÜR MATERIALPRÜFUNGRN DER TECHNIK (CH)	ORGANIS
SVOR		SCHWEIZERISCHE VEREINIGUNG FÜR OPERATIONS RESEARCH (CH)	ORGANIS
SVRS		IMMUNOASSAY SURVEY VALIDATED REFERENCE SERUM	CLINCHEM
SXA		SOFT X-RAY ABSORPTION	METH
SXAPS		SOFT X-RAY APPEARANCE POTENTIAL SPECTROSCOPY	METH
SXE		SOFT X-RAY EMISSION SPECTROSCOPY	METH
SXES	SXE	SOFT X-RAY EMISSION SPECTROSCOPY	METH
SXPES		SOFT X-RAY PHOTO ELECTRON SPECTROSCOPY	
SXPM		SCANNING X-RAY PHOTOELECTRON MICROSCOPY	METH
SXR		SCANNING X-RAY RADIOGRAPHY	METH
SXRA	SYRFA	SYNCHROTON X-RAY FLUORESCENCE ANALYSIS	METH
SXS		SOFT X-RAY SPECTROSCOPY	METH

Abkürzung Akromym abbreviation akronym	Alternative alternative	Bedeutung meaning	Zuordnung related section
2,4,5-T		2,4,5-TRICHLOROPHENOXYACETIC ACID	COMP
T		THYMIDINE	COMP
TA		THERMAL ANAYSIS	METH
TA	TA-LUFT	TECHNISCHE ANLEITUNG LUFT (D)	REG
TA		TECHNISCHER AUSSCHUß (VDI,D)	ORGANIS
TAA		THIOACETAMIDE	COMP
TAAC		TETRA ALKYLAMMONIUM CATION	COMP
TAC		TRIALLYL CYANURATE	COMP
TAC		TECHNISCHER ARBEITSKREIS (BDI,D)	ORGANIS
TAM		4-METHOXY-2-(THIAZOLYL-2-AZO) PHENOL	COMP
TAMA		N-METHYLANILINIUM TRIFLUOROACETATE	COMP
TAMAC		TRIMELLITIC ANHYDRIDE MONOACID	COMP
TAME		N ALPHA-p-TOSYL-L-ARGININE METHYL ESTER	COMP
TAMM		TETRAKIS-(ACETOMERCURI)METHANE	COMP
TAN		THIAZOLYL AZANAPHTHOL	COMP REAG
TAP		TETRA AZA PORPHINATO LIGAND	COMP
TAPA		ALPHA-(2,4,5,7-TETRANITRO-9-FLUORENYLIDENE-AMINOXY) PROPIONIC ACID	COMP
TAPH2		TETRA AZO PORPHORINE	COMP
TAPO		TRIACETYLPHOSPHONIUM OXIDE	COMP
TAPS		3-(TRIS-(HYDROXYMETHYL) METHYLAMINO)-2-HYDROXYPROPANESULFONIC ACID	COMP
TAR		4-(2-THIAZOLYLAZO)-RESORCINE	COMP
TAS	TAS-	TRIS (DIETHYLAMINO)SULFONIUM-LIGAND	COMP
TAT		TUNGSTIC ACID TECHNIQUE	MATER TECHN
TATBA		TRIUMCINOLONE ACETONIDE TERT-BUTYLACETATE	COMP
TATD		THIAMINE 8-(METHYL 6-ACETYLDIHYDROTHIOCTATE) DISULFIDE	COMP
TATM		TRIALLYL TRIMELLITATE	COMP
TB		TEXYLBORANE	COMP
TB		THYMOL BLUE	COMP
TBA	TBAX	TETRABUTYLAMMONIUM SALT (X=B BROMIDE,C CHLORIDE , I IODIDE)	COMP
TBA		THIOBENZANILIDE	COMP
TBA		THIO BARBITURIC ACID	COMP

Abkürzung Akromym abbreviation akronym	Alternative alternative	Bedeutung meaning	Zuordnung related section
TBA		N-TERT-BUTYLACRYLAMIDE	COMP
TBA		THYROXINE BINDING ALBUMINE	CLINCHEM
TBA	2,3,6-TBA	2,3,6-TRICHLOROBENZOIC ACID	COMP
TBA		TRI BROMOANISOLE	COMP
TBAB		TETRABUTYL AMMONIUM BROMIDE	COMP
TBAF		TETRA-n-BUTYLAMMONIUM FLUOROBORATE	COMP
TBAF		TETRABUTYLAMMONIUM FLUORIDE	COMP
TBAH		TETRABUTYLAMMONIUMHYDROXIDE	COMP
TBAHS		TETRA BUTYL AMMONIUM HYDROGENSULFATE	COMP
TBAP		TETRABUTYLAMMONIUM PERCHLORATE	COMP
TBAS		TETRABUTYLAMMONIUM SUCCINIMIDE	COMP
TBBS		2-TERT-BUTYL-BENZOTHIAZOLYL SULPHENAMIDE	COMP
TBC		p-TERT-BUTYLCATECHOL	COMP
TBDA		TEXYLBORANE-N,N-DIETHYLANILINE	COMP
TBDC		BIS-(TETRABUTYLAMMONIUM) DICHROMAT	COMP
TBDMS	TBS-	TERT-BUTYL DIMETHYL SILYL-LIGAND	COMP
TBDMSCI		TERT-BUTYL DIMETHYLSILYL-CHLORIDE	COMP
TBDMSI		1-(TERT-BUTYLDIMETHYLSILYL)-IMIDAZOLE	COMP
TBDMSTF	TBDMSTf	TERT-BUTYL DIMETHYLSILYL TRIFLUOROMETHANESULFONATE	COMP
TBE		TETRABROMOETHANE	COMP
TBF		N-TERT. BUTYLFORMAMIDEN	COMP
TBFBHE		TUBE BUNDLE-FLUIDIZED BED HEAT EXCHANGER	TECHN
TBG		TETRATHIONAT-BRILLIANT GRUEN GALLE	CLINCHEM
TBG		THYROXINE BINDING GLOBULINE	CLINCHEM
TBHC		TERT-BUTYL HYPOCHLORIDE	COMP
TBHP		TERT-BUTYL HYDROPEROXIDE	COMP.
TBHQ		TERT-BUTYL HYDROQUINONE	COMP
TBMP		2-TERT-BUTYL-4-METHYL PHENOL	COMP
TBNE		TERT. BUTYLMETHYLETHER	COMP
TBO		3-((TRIMETHYLSILYL)OXYL)-3-BUTEN-2-ONE	COMP
TBP		2-TERT BUTYL PHENOL	COMP
TBP		TRI-n-BUTYL PHOSPHATE	COMP
TBPA		THYROXINE BINDING PREALBUMINE	CLINCHEM
TBS	TBS-	TERT-BUTYL-DIMETHYLSILYL-LIGAND	COMP
TBS		4-TERT-BUTYLPHENYLSALICYLATE	COMP
TBSCL	TBSI	TERT-BUTYLDIMETHYLSILYL CHLORIDE	COMP

Abkürzung Akromym abbreviation akronym	Alternative alternative	Bedeutung meaning	Zuordnung related section
TBTD		TETRA-BUTYLTHIURAM DISULFIDE	COMP
TBTO		TRI BUTYL TRI OXIDE	COMP
TBTU		TRI-BUTYLTHIOUREA	COMP
TBUP		TRI-n-BUTYLPHOSPHINE	COMP
TC		TOTAL BACTERIAL CONTENT	CLINCHEM
TC		TOTAL CARBON	METH
TC		TECHNICAL COMMITTEE (ISO)	INSTI
TCA		TRICHLOROACETIC ACID	COMP
TCA		2,4,6-TRICHLOROANISOLE	COMP
TCA		THYROCALCITONIN	COMP
TCBOC	TCBOC-Cl	CHLOROFORMICACID-2,2,2-TRICHLOR-TERT-BUTYLESTER	COMP REAG
TCC		TRANSPORT CONDUCTIV COATING	TECHN
TCCH		TETRACHLOROCYCLOHEXANE	COMP
TCD	WLD, HWD	THERMAL CONDUCTIVITY DETECTOR	INSTR
TCD	TCD-REAG	2,2,5,5-TETRAKIS(CARBOXYMETHYLTHIO)1,4-DITHIANEW	COMP REG
TCDD	PCDD4	TETRACHLORO-p-DIBENZO DIOXIN	COMP
TCDF	PCDF4	TETRACHLORO DIBENZO FURAN	COMP
TCDNB		TETRACHLORO DINITROBENZENE	COMP
TCDS		TANDEM CYLINDRICALDEFLECTOR SPECTROMETRY	METH
TCE		TRICHLORO ETHYLENE	COMP
TCE		TRICHLOROETHANOL	COMP
TCE	Tce-	2,2,2-TRICHLOROETHYL-LIGAND	COMP
TCEP		1,2,3-TRIS-(2-CYANOETHOXY) PROPANE	MATER
TCHQ-DE		TETRACHLORO HYDROQUINONE DIALKYLETHER	COMP
TCL	TCl	TEREPHTHALOYL CHLORIDE	COMP
TCM		TRICHLOROMELAMINE	MATER POYM
TCN		TETRACHLORO NAPHTHALENE	COMP
TCNB		TETRACHLORO NITROBENZENE	COMP
TCNE		TETRACYANO ETHYLENE	COMP
TCNP		TRICYCLIC NUCLEOSIDE 5'-PHOSPHATE	COMP CLINCHEM
TCNQ		7,7,8,8-TETRACYANO QUINODIMETHANE	COMP
TCP		TRICHLOROPHENOL (DIV SUBTITUT.)	COMP
TCP		TRI CRESYL PHOSPHATE	COMP
TCPE	2,4,5-TCPE	2,4,5-THRICHLORO PHENOXY ETHANOL	COMP
TCPO		1,2-EPOXY-3,3,3-TRICHLOROPROPENE	COMP

Abkürzung Akromym abbreviation akronym	Alternative alternative	Bedeutung meaning	Zuordnung related section
TCPO		BIS-2,4,6-TRICHLORO PHENYL OXALATE	COMP
TCQ		N-2,6-TRICHLOROBENZOQUINONE IMINE	COMP
TCTFP		1,1,2,2-TETRACHLORO-3,3,4,4-TERAFLUOROCYCLOBUTANE	COMP
TDB		TRACKING DYE BOUNDARY	TECHN
TDBA	TDBA-	TETRADECYL DIMETHYLBENZYL AMMONIUM CATION	COMP
TDC		TIME TO DIGITAL CONVERTER	INSTR
TDD		TOTAL DAILY DIET	CLINCHEM
TDD		TWO DETECTOR DELAY	INSTR
TDDA		9,11-TETRADECADIEN-1-OL ACETATE	COMP
TDE	DDD	TETRACHLORO DIPHENYL DICHLOROETHANE (DIV.)	COMP
TDE		2,2'-(2,5,8,11-TETRAOXA-1,12-DODECANEDIYL)BISOXIRANE	COMP
TDI		TOLUENE-2,4-DIISOCYANATE	COMP
TDID		DIMERIC TOLUYLENE DIISOCYANATE	COMP
TDIU	2,4-TDIU	2,4-(1-TOLYL-DI-(3-n PROPYL-3(4 NITROBENZL)) UREA	COMP
TDM		THERAPEUTIC DRUG MONITORING	CLINCHEM
TDMS		THERMAL DESORPTION MASS SPECTROMETRY	METH
TDN		1,1,6-TRIMENTHYL DIHYDRO NAPHTHALENE	COMP
TDNA	T-DNA	TUMOR DESOXYRIBONUCLEIC ACID	CLINCHEM
TDP	TDP-	TETRA DECYL PHOSPHONIUM-CATION	COMP
TDP		THYMIDINE DIPHOSPHATE	CLINCHEM COMP
TDS		THERMAL DESORPTION SPECTROSCOPY	METH
TDS		TOTAL DISSOLVED SOLIDS	METH
TDS		THERMAL DIFFUSE SCATTERING	METH
TDS		THERMAL DESORPTION SPECTROSCOPY	METH
TDT	TdT	TERMINALE DESOXY NUCLEOSIDYL TRANSFERASE	CLINCHEM
TE	PD	TEILENDLADUNG	TECHN
TEA		TRIETHANOLAMINE	COMP
TEA		TRI ETHYLALUMINIUM	COMP
TEA		THERMAL ENERGY ANALYZER	INSTR

Abkürzung Akromym abbreviation akronym	Alternative alternative	Bedeutung meaning	Zuordnung related section
TEA		TRANSVERSE-EXCITED ATMOSPHERIC PRESSURE (LASER)	INSTR
TEA	T.E.A.	TETRAETHYLAMMONIUM -CATION	COMP
TEAB		TETRAETHYLAMMONIUM BICARBONATE	COMP
TEAC	TEAC-	TETRAETHYL AMMONIUM CATION	COMP
TEAE-C	TEAE-CELL	TRIEHYLAMINOETHYL CELLULOSE	MATER
TEAP		TETRAETHYLAMMONIUM PERCHLORATE	COMP
TEAS		TETRAETHYLAMMONIUM SUCCINIMIDE	COMP
TEBA		TRIETHYL BENZYL AMMONIUM SALT	COMP
TED		TRANSMISSION ELECTRON DIFFRACTION	METH
TED	DABCO,TEM	TRIETHYLENE DIAMINE (=1,4-DIAZABICYCLO(2.2.2.)OCTANE)	COMP
TEDP		TETRAETHYL DITHIOPYROPHOSPHORSAEURE ESTER	COMP
TEED		N,N,N',N'-TETRAETHYL ETHYLENEDIAMINE	COMP
TEELS		TRANSMISSION ELECTRON ENERGY LOSS SPECTROSCOPY	METH
TEF	TUF	TRUEBUNGSEINHEITEN FORMAZIN	METH
TEG		TRACE ELEMENTS IN GLAS	MATER
TEG		TOP ENTRY GONIOMETER	INSTR
TEG		TRIETHYLENE GLYCOL	COMP
TEGDA		TRIETHYLENE GLYCOL DIACETATE	COMP
TEGEWA		VERBAND DER TEXTILHILFSMITTEL,GERBSTOFF & WASCHROHSTOFF INDUSTRIE (D	ORGANIS
TEL		TETRAETHYL LEAD	COMP
TELS		TRANSMISSION ENGERGY LOSS SPECTROSCOPY	METH
TELSCA		TRANSMISSION ENERGY LOSS ELECTRON SPECTROSCOPY FOR CHEMICAL ANAYSIS	METH
TEM		TRANSMISSION ELECTRON MICROSCOPY	METH
TEM	TED,DABCO	TRI ETHYLENE MELAMINE =1,4-DIAZABICYCLO-(2.2.2)OCTANE	COMP
TEMPO		2,2,6,6-TETRAMETHYL PIPERIDINO OXY- FREE RADICAL	COMP
TEPA		TETRAETHYLENE PENTAMINE	COMP
TEPP		TETRAETHYLPYROPHOSPHATE	COMP
TEPS	TEP's	TRIAL ENVIRONMENTAL POOLS	ENVIR
TES		N,N,N',N'-TETRAETHYLSULFAMIDE	COMP

Abkürzung Akromym / abbreviation akronym	Alternative / alternative	Bedeutung / meaning	Zuordnung / related section
TES		2-(TRIS(HYDROXYMETHYL)METHYLAMINO)-1-ETHANESULFONIC ACID	COMP
TES	TES-	TRIETHYLSILYL-LIGAND & TRIETHYLSILANE	COMP
TET		TRIETHYLENE TETRAMINE	COMP
TETA	TETA 4-	1,4,8,11-TETRAAZACYCLOTETRADECANE-N,N',N'',N'''-TETRAACETATE	COMP
TETA		TRIETHYLENE TETRAMINE	COMP
TETD		TETRAETHYL THIURAM DISULFIDE	COMP
TETM		TETRAETHYLTHIURAM MONOSULFIDE	COMP
TETN		TRIETHYL AMINE	COMP
TEU		TETRAETHYL UREA	COMP
TF		TRANSFER FUNCTION	THEOR
TF		TRANS SCRIPTION FACTOR	CLINCHEM
TFA		TRIFLUOROACETYLACETONE	COMP REAG
TFA	TFA-	TRIFLUOROACETYL-LIGAND & ACETATE	COMP
TFA		THIN FILM ANALYSIS	METH
TFA		TARGET TRANSFORMATION FACTOR ANALYSIS	EVAL THEOR
TFAA		TRIFLUORO ACETIC ANHYDRIDE	COMP
TFAI		TRIFLUOROACETYL-IMIDAZOLE	COMP
TFAME	TFA-Me	METHYL TRIFLUORO ACETATE	COMP
TFE		2,2,2-TRIFLUOROETHANOL	COMP
TFE	TFE-	TRIFLUOROETHYL-LIGAND	COMP
TFM		4-NITRO-3-(TRIFLUOROMETHYL) PHENOL	COMP
TFMC	TFMC-RE	TRIS(3-(TRIFLUOROMETHYL-HYDROXYMETHYLENE)d-CAMPHORATO) -RARE EARTH	COMP
TFMC		THIN FILM MERCURY ELECTRODE	INSTR
TFMSA		TRIFLUOROMETHANE SULFONIC ACID	COMP REAG
TFP		TRANS FIBRE PHOTOGRAPHY	METH
TFT		THIN FILM TRANSISTOR	ELECT
TFTR		TOKAMAK FUSSION TEST REACTOR	TECHN
TFTR		TRITIUM FISSION TEST REACTOR	TECHN
TGA		THERMOGRAVIMETRIC ANALYSIS	METH
TGG		TERBIUM-GALLIUM-GARNET (GRANAT)	MATER
TGL		THERMAL GRADIENT LAMP	INSTR
TGM	TGMV	TOMATO GOLDEN MOSAIC VIRUS	CLINCHEM
TGMDA		TETRAGLYCIDYL 4,4'-METHYLENDIANILIN	COMP
TGS		TRIGLYCINSULFATE	COMP INSTR

Abkürzung Akronym	Alternative	Bedeutung	Zuordnung
abbreviation akronym	alternative	meaning	related section
TH		TECHNISCHE HOCHSCHULE (D)	INSTI
THAM	TRIS	ALPHA-TRIS-(HYDROXXYMETHYL)-METHYLAMINE	COMP
THBC		1,2,34,-TETRAHYDRO-BETA-CARBOLINES	CLINCHEM
THBP		2',4',5'-TRIHYDROXY-BUTYROPHENONE	COMP
THC	THC's	6a-7,8,10a- TETRAHYDROCANNABIOL(S)	CLINCHEM COMP
THC		TETRAHYDROCORTISONE	COMP CLINCHEM
THEED		TETRAHYDROXYETHYL ETHYLENE DIAMINE	COMP
THEED		TRANSMISSION HIGH ENERGY SCANNING ELECTRON DIFFRACTION	METH
THF		TETRAHYDROFOLIC ACID	COMP
THF		TETRAHYDROFURAN	COMP
THFA		TETRAHYDROFURFURYL ALCOHOL	COMP
THFC	THFC-RA	TRIS (3-HEPTAFLUOROPROPYLHYDROXY-METHYLENE) -d-CAMPHORATO RARE EARTH	COMP
THI		1-METHYL-3,4,5-TRIHYDROXY INDOLE	COMP
THIP		4,5,6,7-TETRAHYDROISOXAZOLO(5.4-c)-PYRIMIDIN-3(2H)-ONE	COMP
THM		TRIHALOMETHANE	COMP
THP		TETRAHYDROPYRAN OR TETRAYDROPYRANYL	COMP
THP		TETRAHYDRO PAPAVEROLINE	COMP
THQ		2,3,5,6-TETRAHYDROXY-2,5-BENZOQUINONE (=TETROQUINONE)	COMP
THR	Thr	THREONINE	COMP
THT		TETRAHYDROTHIOPHENE	COMP
THTR		THORIUM HOCHTEMPERATUR REAKTOR (D)	TECHN
THYRISTOR		THYRATRON TRANSISTOR	ELECT
TI		TEXAS INSTRUMENTS	COMPANY
TIAFT		THE INTERNATIONAL ASSOCIATION OF FORENSIC TOXICOLOGIST(S)	ORGANIS
TIBA		TRI IODOBENZOIC ACID	COMP
TIBA		TRIISOBUTYL ALUMINIUM	COMP
TIC		TOTAL ION CURRENT	METH
TIC		TOTAL ION CHROMATOGRAPHY	METH
TIC		TOTAL INORGANIC CARBON	MATER
TICP		TOTAL ION CURRENT PROFILE	METH
TID		THERMAL IONIZATION DETECTOR	INSTR
TIM		TRIOSEPHOSPHATE ISOMERASE	CLINCHEM
TIM		TOTAL ION MONITORING	INSTR

Abkürzung Akromym abbreviation akronym	Alternative alternative	Bedeutung meaning	Zuordnung related section
TIMS		THERMAL IONIZATION MASS SPECTROMETRY	METH
TIPSCI		1,3-DICHLORO-1,1,3,3-TETRAISOPROPYL-DISILOXANE	COMP
TISAB		TOTAL IONIC STRENGTH ADJUSTMENT(ADJUSTING) BUFFER	COMP
TITC		TSCA INTERAGENCY TESTING COMMITTEE (US)	INSTI
TK		THYMIDINE KINASE	CLINCHEM
TKN		TOTAL KJELDAHL NITROGEN	COMP
TKP	TCP	TRI KRESYL PHOSPHATE	COMP
TL		THERMOLUMINESCENCE	METH
TLA		THIN LAYER ACTIVATION	TECHN
TLC	DC	THINLAYER CHROMATOGRAPHY	METH
TLCF		THIN LAYER CHROMATOGRAPHY COMBINED WITH FLUORIMETRY	METH
TLCK		7-AMINO-1-CHLOR-3-(p-TOYLAMIDO)-HEPTAN-2-KETON	COMP
TLD		THERMOLUMINESZENZDOSIMETER	INSTR
TLGPC		THINLAYER GEL PERMEATION CHROMATOGRAPHY	METH
TLM		TRANSMISSION LIGHT MICROSCOPE	INSTR
TLM	TLm	MEDIUM TOLERANCE LIMIT	CLINCHEM
TLV		THRESHOLD LIMIT VALUES	EVAL REG
TLV-STEL		TRESHOLD LIMIT VALUES SHORT TERM EXPOSURE LIMIT	REG
TLV-TWA		TRESHOLD LIMIT VALUES TIME WEIGHTED AVERAGE	REG
TLVC	TLV-C	THRESHOLD LIMTIT VALUES CEILING	REG
TM		TRADE MARK	REG
TMA		THERMO MECHANICAL ANALYSIS	METH
TMA		TRIMETHOXY AMPHETAMINE	COMP
TMA		TETRAMETHYL AMMONIUM ION	COMP
TMAC		TRIMELLITIC ANHYDRIDE MONOACID CHLORIDE	COMP
TMAEMC		2-TRIMETHYL AMMONIUMETHYLMETHACRYLIC CHLORIDE	COMP
TMAH		TETRAMETHYL AMMONIUM HYDROXIDE	COMP
TMAH		TRIMETHYL ANILINIU HYDROXIDE	COMP
TMAO		TRIMETHYL AMIN-N-OXID	COMP
TMAT		TETRAMETHYLAMMONIUM TRIBROMIDE	COMP

abbreviation akronym	alternative	meaning	related section
TMAT		TRIS-2,4,6-(1-(2-METHYL)AZIRIDINYL)-1,3,5-TRIAZINE	COMP
TMB		TETRAMETHYL BENZIDINE (3,3',5,5' OR N,N,N',N')	COMP
TMB		TRIMETHYL BROMO SILANE	COMP
TMB		3,3',5,5'-TETRAMETHYL BENZIDINE	COMP
TMBA		3,4,5-TRIMETHYLBENZALDEHYDE	COMP
TMC		THICK MOULDING COMPONENT	MATER
TMC		3,3,5-TRIMETHYL CYCLOHEXANOL	COMP
TMC1		TAURUS MOLECULAR CLOUD	ENVIR
TMCPD		AMINO-2-AMINOMETHYL-3.3.5(3.5.5)TRIMETHYLCYCLOPENTAN	COMP
TMCS		TRIMETHYLCHLOROSILANE	COMP
TMD		TETRAGYCIDYLE METHYLENE DIANILINE	COMP
TMD		2,2,4(2,4,4)-TRIMETHYL-HEXAMETHYLENE DIAMINE	COMP
TMD		1,1,10-TRIMETHYL-TRANS-2-DECALOL	COMP
TMDI		2.2.4(2.4.4)TRIMETHYL-HEXAMETHYLENE DIISOCYANATE	COMP
TMDS		1,1,3,3-TETRAMETHYL DISILAZANE	COMP
TME		(N',N''-TETRAMETHYL-3,8-DIAMINO)1-PHENYL-N-ETHYLPHENANTHRIDIUM BROMIDE	COMP
TME		TRIMETHYL ETHIDINIUM-ION	COMP
TMEDA	TMED	N,N,N',N'-TETRAMETHYLENE DIAMINE	COMP
TMETD		TETRAMETHYL TETRAETHYL THIURAM DISULFIDE	COMP
TMG		METHYL BETA-D-THIOGALACTOSIDE	COMP
TMIC	TOSMIC	TOSYLMETHYL ISOCYANIDE	COMP
TMK		4,4' BIS(DIMETHYLAMINO)THIOBENZOPHENONE =THIO MICHERS KETON	COMP
TMM		TRIMETHYLENE METHANE	COMP
TMO		TRIMETHYLAMINE N-OXIDE	COMP
TMP		TRIMETHYLPHENOL (DIVERSE)	COMP
TMP		2,2,6,6-TETRAMETHLPIPERIDINE	COMP
TMP		THYMIDINE 5'-MONOPHOSPHATE	COMP
TMP		TRIMETHYLOL PROPANE	COMP
TMPTMA		TRIMETHYLOLPROPANE TRIACETACRYLATE	COMP
TMS	TMS-	TRIMETHYLSILYL-LIGAND	COMP

Abkürzung Akromym abbreviation akronym	Alternative alternative	Bedeutung meaning	Zuordnung related section
TMS	TIMS	THERMIONIC MASS SPECTROMETY	METH
TMSA		TRIMETHYLSILYL AMIDE	COMP
TMSAN		TRIMETHYL SILYL ACETONITRILE	COMP
TMSDEA		N,N-DIETHYL-1,1,1-TRIMETHYLSILYLAMINE	COMP
TMSI	TSI TSIM	TRIMETHYLSILYL IMIDAZOLE	COMP
TMTD		TETRAMETHYL THIURAM DISULFIDE	COMP
TMTFTH		TRIMETHYL-(ALPHA-TRIFLOURO- - TOLYL)AMMONIUM HYDROXIDE	COMP
TMTM		TETRAMETHYL THIURAM MONOSULFIDE	COMP
TMTT		TETRAMETHYL THIURAM TETRASULFIDE	COMP
TMTU		TETRAMETHYL THIOUREA	COMP
TMU		TETRAMETHYL UREA	COMP
TMV		TABAK MOSAIK VIRUS	CLINCHEM
TN	TNC	TOTAL NITROGEN CONTENT	METH
TNAA		THERMAL NEUTRONS ATOMIC ABSORPTION	METH
TNBA		TRI-n-BUTYLALUMINIUM	COMP
TNBS		2,4,6-TRINITRO BENZENE SULFONICACID	COMP
TNBT		TETRANITRO BLUE TETRAZOLIUM	COMP
TNC		TWISTED NEMATIC CELLS	TECHN
TNF		2,4,7-TRINITROFLUORENONE	COMP
TNF		TUMOR NECROSIS FACTOR	CLINCHEM
TNM		TETRANITROMETHANE	COMP
TNO	8-TNO	8-THIOL CHINOLIN-N-OXIDE	COMP REAG
TNO		ORGANISATIE VOOR TOEGEPAST NATUURWE-TENSCHAPPELY ONDERZOEK (DELFT,NL)	ORGANIS
TNPA		TRI-n-PROPYLALUMINIUM	COMP
TNPP		TRINONYL PHENYLPHOSPHATE	COMP
TNS		6-(p-TOLUIDINO)-2-NAPHTHALENE-SULFONICACID , MOST ALKALISALT	COMP
TNT		2,4,6-TRINITROTOLUENE	COMP
TOA		TERT.-OCTYLACRYLAMIDE	COMP
TOBO		2-(p-TOLYL) BEZOXAZOLE	COMP
TOC		TOTAL ORGANIC CARBON	METH COMP
TOCL	TOCl	TOTAL ORGANIC CHLORINE	METH COMP
TOCP		TRI ORTHO CRESYL PHOSPHATE	COMP
TOF		TIME OF FLIGHT	INSTR
TOF		TRIOCTYLPHOSPHATE	COMP

Abkürzung Akromym / abbreviation akronym	Alternative / alternative	Bedeutung / meaning	Zuordnung / related section
TOF-MS		TIME OF FLIGHT MASS SPECTROMETRY	METH
TOFA		TIME OF FLIGHT ANALYSIS	METH
TOLZ		THIRD ORDER LAUE ZONE	METH
TOMAL		TASK ORIENTED MICROPROCESSOR APPLICATION LANGUAGE	ELECT
TOPO		TRI-n-OCTYL PHOSPHINE OXIDE	COMP
TOSCA	TSCA	TOXIC SUBSTANCES CONTROL ACT (US)	REG
TOSMIC		(p-TOLUENESULFONYL)METHYL ISOCYANIDE	COMP
TOX		TOTAL ORGANIC HALOGEN	METH
TOXREG		TOXICITY REGULATIONS (FRANKLIN INSTITUTE,US)	LIT
TP	TPP	TRIPHENYLPHOSPHATE	COMP
TP		TECHNISCHES PRODUKT	MATER
TP		THYMOLPHTHALEIN	COMP
TPA		TEREPHTHALIC ACID	COMP
TPA		TETRAPROPYLAMMONIUM-LIGAND	COMP
TPA		12-O-TETRADECANONYLPHORBOL-13-ACETATE	COMP
TPA	TPA-	TETRAPHENYL ARSONIUM CATION	COMP
TPA		TISSUE POLYPEPTIDE ANTIGENE	CLINCHEM
TPA		TWO PHOTON ABSORPTION	METH
TPB		1,1,4,4-TETRAPHENYLBUTADIENE	COMP
TPC		THYMOLPHTHALEIN COMPLEXONE	COMP
TPC		THREE PHASES CATALYSIS	METH
TPC		TOTALLY PYROLYTIC CUVETTE (AAS)	INSTR
TPCD		TETRAPHENYLCYCLOPENTADIENONE	COMP
TPCK		L-1-P-TOSYLAMINO-PHENYLETHYLCHLOROMETHYL KETONE	COMP
TPD	DTPT	THIAMINE PROPYL DISULFIDE	COMP
TPD		TEMPERATURE PROGRAMMED DESORPTION	METH
TPE		TETRAPHENYLETHYLENE	COMP
TPE		TWO PHOTON EXCITATION	METH
TPE	TPI	THERMOPLASTISCHES ELASTOMER	MATER
TPEA	TPeA	TETRAPENTYL AMMONIUM HALIDE	COMP
TPEF		TWO PHOTON EXCITED FLUORESCENCE	METH
TPF		TRIPHENYLPHOSPHATE	COMP
TPF		TWO PHOTON FLUORESCENCE	METH
TPGC		TEMPERATURE PROGRAMMED (PROGRAMMING) GASCHROMATOGRAPHY	METH
TPI	TPE	THERMOPLASTOID	MATER

Abkürzung Akromym abbreviation akronym	Alternative alternative	Bedeutung meaning	Zuordnung related section
TPM		TOTAL PARTICULATE MATTER	METH
TPMP	TPMP-	TRIPHENYLMETHYLPHOSPHONIUM-ION	COMP
TPMP		TETRADIMETHYL-AMINOPHENYLPENTAMETHIN PERCHLORAT	COMP
TPN	NADP	TRIPHOSPHOPYRIDINE NUCLEOTIDE, SODIUM SALT	COMP
TPNH	NADH	REDUCED TRIPHOSPHOPYRIDINE NUCLEOTIDE, SODIUM SALT	COMP
TPO		THERMOPLASTIC OLEFINS	MATER POLYM
TPP	TP	TRIPHENYL PHOSPHATE	COMP
TPP		TETRAPHENYLPORPHYRINE	COMP
TPP		TRIS-(ORTHO-PHENYLEN-DIOXY)CYCLOTRIPHOSPHAZEN	MATER
TPP		THIAMIN PYROPHOSPHATE	CLINCHEM
TPPH2		TETRAPHENYL PORPHINE	COMP
TPS		TRIPHENYLSULFONIUM CHLORIDE	COMP
TPS	TPS-	2,4,6-TRIISOPROPYLBENZENESULFONYL-LIGAND	COMP
TPSH		2,4,6-TRIISOPROPYL-BENZENE-SULFONIC ACID HYDRAZIDE	COMP
TPTD		TETRA ISOPROPYL THIURAM DISULFIDE	COMP
TPTZ		2,4,6-TRIS (2'-PYRIDYL)-S-TRIAZINE	COMP
TPTZ	TTC	2,3,5-TRIPHENYL-2H-TETRAZOLIUM CHLORIDE	COMP
TPU		THERMOPLASTIC UREA	MATER
TPZ		2,5-DIPHENYL-3-(4-STYRYLPHENYL)TETRAZOLIUM CHLORIDE	COMP
TQMS		TRIPLE QUADRUPOLE MASS SPECTROMETRY	METH
TRA		TECHNICAL AND REGULATORY AFFAIRS (CESIO)	ORGANIS
TRB		TECHNISCHE REGELN DRUCKBEHÄLTER	REG
TRBDD	PBDD3	TRIBROMO-p-DIBENZODIOXIN	COMP
TRBDF	PBDF3	TRIBROMO DIBENZOFURANE	COMP
TRBF	TRbF	TECHNISCHE REGELN BRENNBARE FLUESSIGKEITEN (D)	REG
TRCBZ	TrCBZ	TRICHLOROBENZENE	COMP
TRCDD	PCDD3	TRICHLORO-p-DIBENZODIOXINE	COMP
TRCDF	PCDF3	TRICHLORO-DIBENZOFURAN	COMP
TRCP	TrCP	TRICHLOROPHENOL	COMP
TRF		TECHNISCHE REGELN FLÜSSIGGAS (D)	REG
TRF		THYROTROPIN RELEASING FACTOR	CLINCHEM

Abkürzung Akromym	Alternative	Bedeutung	Zuordnung
abbreviation akronym	alternative	meaning	related section
TRFA	TXRF	TOTALREFLEXIONS RÖNTGEN FLUORESZENZ ANALYSE	METH
TRFIA	TR-FIA	TIME RESOLVED FLUORESCENT IMMINO ASSAY	CLINCHEM
TRGI		TECHNISCHE REGELN FÜR GAS INSTALLATION (D)	REG
TRH		THYREO RELEASING HORMONE =TSH-RELEASING HORMONE	CLINCHEM
TRI		TRICHLOROETHYLENE	COMP
TRI		TEXTILE RESEARCH INSTITUTE (S)	INSTI
TRIAC	THYRISTOR	TRIODE ALTERNATING CURRENT SEMICONDUCTOR SWITCH	ELECT
TRICINE	Tricine	N-(tris (TRIHYDROXYMETHYL) GLYCINE	COMP
TRICT	TRITC	TETRAMETHYLRHODAMINE ISOTHIOCYANATE	COMP
TRIGLYME	Triglyme	TRIETHYLENE GLYCOL DIMETHYLETHER	COMP
TRIM		TRIMETHYLOLE PROPANE TRIMETHACRYLATE	MATER POLYM
TRIS	THAM	TRIS (HYDROXYMETHYL) AMINOMETHANE	COMP
TRITC	TRICT	TETRAMETHYL RHODAMINE ISOTHIOCYANATE	COMP
TRIX		TOTAL RATE IMAGING OF X-RAYS	METH
TRK		TECHNISCHE RICHTKONZENTRATION (D)	REG
TRP	Trp	TRYPTOPHANE	COMP
TRPGDA		TRIPROPYLENEGLYCOL DIACRYLATE	COMP
TRRFA	TR-RFA	TIME RESOLVED X-RAY FLUORESCENCE ANALYSIS	METH
TRRR		TIME RESOLVED RESONANCE RAMAN SPECTROMETRY	METH
TRRS		TIME RESOLVED RAMAN SPECTROSCOPY	METH
TRT		TRENNUNG-REAKTION-TRENNUNG (DC-TECHN)	METH
TSA		TOTAL SURFACE AREA	METH
TSC		TROCKENSAEULENCHROMATOGRAPHIE	METH
TSC		TRYPTOSE-SULFIT-CYCLOSERIN	CLINCHEM
TSC		THIOAUTAZON	COMP
TSCI		TOXIC SUBSTANCES CONTROL INSTITUTE (US)	INSTI
TSD		THERMIONIC SPECIFIC DETECTOR	INSTR
TSDA		TANDEM SPHERICAL DEFLECTOR ANALYSER	INSTR
TSED		TRANSMISSION SCANNING ELECTRON DIFFRACTION	METH

abbreviation akronym	alternative	meaning	related section
TSEE		THERMALLY STIMULATED EXOELECTRON EMISSION	METH
TSEM		TRANSMISSION SCANNING ELECTRON MICROSCOPY	METH
TSH		THYROID STIMULATING HORMONE =THYOTROPIC HORMONE	CLINCHEM
TSH		TOLUENE SULFONYL HYDRAZIDE	COMP POLYM
TSI	TMSI	N-TRIMETHYLSILYL IMIDAZOLE	COMP REAG
TSIM	TSI	N-TRIMETHYL SILYL IMIDAZOLE	COMP
TSL		TRISTATE LOGIC	ELECT
TSNI		1-(p-TOLUENESULFONYL)-4-NITROIMIDAZOLE	COMP
TSO		TRANS STILBENE OXIDE	COMP
TSP		TRIPLE SUPERPHOSPHATE	MATER
TSP		THROMBOSPONDIN	CLINCHEM
TSPP		TETRA SODIUM PYROPHOSPHATE	COMP
TSQ		TRIPLE STAGE QUADRUPOLE MASS SPECTROMETRY	METH
TSSC		TOXIC SUBSTANCES STRATEGY COMMITTEE (US)	INSTI
TSSG		TOXIC SUBSTANCES STRATEGY GROUP (US)	INSTI
TTA		TOTYLTRIAZOLE	COMP
TTA		1-(THENOYL-(2'))-3.3.3-TRIFLUOROACETON	COMP
TTC		TRITHIOCARBONATE	COMP
TTC	TPTZ	2,3,5-TRIPHENYLTETRAZOLIUM CHLORIDE	COMP
TTCA		THIAZOLIDINE-2-THIONE-CARBOXYLIC ACID	COMP
TTD		TETRAETHYL THIOPEROXY DICARBONIC DIAMIDE	COMP
TTEGDA		TETRAETHYLENE GLYCOL DIACRYLATE	COMP
TTF		TETRATHIAFULVALENE	COMP
TTFA		THALLIUM (III) TRIFLUOROACETATE	COMP
TTHA		TRIETHYLENE TETRAAMINE HEXAACETIC ACID	COMP
TTI		TRANSMITTER TERMINAL IDENTIFICATION	ELECT
TTL		TRANSISTOR-TRANSISTOR LOGIC	ELECT
TTN		THALLIUM THREE NITRATE	COMP
TTP		THYMIDINE-5'-TRIPHOSPHATE	COMP
TTS		TRANSDERMALES THERAPEUTISCHES SYSTEM	CLINCHEM
TTT		TETRATHIOTETRAZEN -LIGAND	COMP
TTT		TRIETHYL TRIMETHYLENE TRIAMINE	COMP

Abkürzung Akromym abbreviation akronym	Alternative alternative	Bedeutung meaning	Zuordnung related section
TTTL		MODIFIED TRANSISTOR TRANSISTOR LOGIC	ELECT
TTV		TECHNISCHE TANKVORSCHRIFTEN (CH)	REG TECHN
TTX		TETRODOTOXIN	CLINCHEM
TU		TECHNISCHE UNIVERSITAET (D)	INSTI
TU		THIOUREA	COMP
TUF	TEF	TURBIDITY UNITS FORMAZINE	METH
TUIS		TRANSPORT UNFALL INFORMATIONS SYSTEM (VCI,D)	TECHN
TUV		THERMAL ULTRAVIOLET	METH
TVB-N		TOTAL VOLATILE BASIC NITROGEN	METH
TVC		THERMAL TO VOLTAGE CONVERTER	INSTR
TVI		TEXTIL VEREDELUNGSINDUSTRIE (D)	ORGANIS
TVO		TRINKWASSER VERORDNUNG (D)	REG
TWA		8-HOUR TIME-WEIGHTED AVERAGE	METH EVAL
TX	TXA2,TXB2	THROMBOXANE div	CLINCHEM
TXP		TRI-2,4-XYLENYL PHOSPHATE	MATER
TYR	Tyr	TYROSINE	COMP
TZ		TRENNZAHL	EVAL
TZ		THROMBINZEIT	CLINCHEM
TÜRK KOD.		TÜRK KODEKSI (TR)	NORM
TÜV		TECHNISCHER ÜBERWACHUNGSVEREIN (D)	COMPANY

Abkürzung Akromym abbreviation akronym	Alternative alternative	Bedeutung meaning	Zuordnung related section
UAI		UNION DES ASSOCIATIONS INTERNATIONALES (BRÜSSEL)	ORGANIS
UBA		UMWELTBUNDESAMT (BERLIN,D)	INSTI
UCG		URIN CHORIONGONATROPIN	CLINCHEM
UDIRL		UNIVERSITY OF DURHAM INDUSTRIAL RESEARCH LABORATORIES	INSTI
UDMH		UNSYMMMETRIC DIMETHYLHYDRAZINE	COMP
UDP		URIDINE 5'-DIPHOSPHATE	COMP
UDPG		URIDINE-5'-DIPHOSPHOGLUCOSE DINATRIUMSALZ	CLINCHEM
UDPGA		URIDINE-5'-DIPHOSPHOGLUCURONIC ACID	COMP
UEATC	UEAtc	UNION EUROPEENNE POUR L'AGREMENT TECHNIQUE DANS LA CONSTRUCTION	ORGANIS
UEG		UNTERE EXPLOSIONSGRENZE	TECHN
UF		UREA FORMALDEHYDE RESIN	MATER POLYM
UFL		UPPER FLAMMABLE LIMIT	TECHN
UFR		URINARY FLOW RATE	CLINCHEM
UFS		UREA FORMALDEHYDE FOAM	MATER POLYM
UGC		UNIVERSITY GRANTS COMMITTEE (GB)	INSTI
UHF		UTRA HIGH FREQUENCY	INSTR
UHF		UNRESTRICTED HARTREE FOCK	THEOR
UHMPE	UHMWPE	ULTRAHIGH MOLECULARWEIGHT POLYETHYLENE	MATER
UHPR		ULTRA HIGH DEPTH RESOLUTION	INSTR
UHT		ULTRA HIGH TEMPERATURE	TECHN
UHV		ULTRA HIGH VACUUM	INSTR
UHVEM		ULTRA HIGH VOLTAGE ELECTRON MIROSCOPY	METH
UIC		UNION DES INDUSTRIES CHIMIQUES (F)	ORGANIS
UICC		UNION INTERNATIONAL CONTRE LE CANCER (INT)	ORGANIS
UIPP		UNION DES INDUSTRIES DE LA PROTECTION DES PLANTES	ORGANIS
UKCIS		UNITED KINGDOM CHEMICAL INFORMATION SERVICE (GB)	LIT
UKEA		UNITED KINGDOM ATOMIC ENERGY AGENCY (GB)	INSTI
ULIDAT		UMWELT LITERATUR DATENBANK (D)	LIT
ULPA		ULTRALOW PENETRATION AIR FILTER	TECHN

Abkürzung Akromym / abbreviation akronym	Alternative / alternative	Bedeutung / meaning	Zuordnung / related section
ULV		ULTRA LOW VOLUME SPRAYING (PFANZENSCHUTZ)	TECHN
UMK		UMWELT MINISTER KONFERENZ (D)	INSTI
UMP		URIDINE 5'-MONOPHOSPHATE	COMP
UMPA		UNIVERSAL MICROPROBE MASS ANALYSER	INSTR
UMPLIS		UMWELT PLANUNGS INFORMATIONSSYSTEM (D)	LIT ENVIR
UNCSTD		UN CONDERENCE ON SCIENCE AND TECHNOLOGY FOR DEVELOPMENT	CONF
UNCTAD		UNITED NATIONS CONFERENCE OF TRADE AND DEVELOPMENT	CONF
UNDP		UNITED NATIONS DEVELOPMENT PROGRAMME	
UNEP		UNITED NATIONS ENVIRONMENT PROGRAM (UN , INT)	ENVIR
UNEPA		UNITED NATIONS ENVIRONMENTAL PROTECTION AGENCY	INSTI
UNESCO		UNITED NATIONS EDUCATIONAL,SCIENTIFIC AND CULTURAL ORGANIZATION	ORGANIS
UNICE		UNION DES INDUSTRIES DE LA COMMUNAUTE EUROPEENNE	ORGANIS
UNICELPE		ASSOC.EUROPEENNE DES PRODUCTEURS DE PROTEINES UNICELLULAIRERS	ORGANIS
UNIDO		UNITED NATIONS INDUSTRIAL DEVELOPMENT ORGANIZATION (INT)	ORGANIS
UNIEO		UNITED NATIONS INDUSTRY AND ENVIRONMENTAL OFFICE	INSTI
UNILAC		UNIVERSAL ION LINEAR ACCELERATOR (DARMSTADT,D)	INSTI
UNSF		UNITED NATIONS SPECIAL FUND	INSTI
UO	RO	UMKEHR OSMOSE	METH
UOP		UNIVERSAL OIL PRODUCTS	MATER
UP		UNKNOWN PATTERN	EVAL
UP		UNSATURATED POLYESTER	MATER
UPA	UpA	URIDYLYL (3'-5') ADENOSINE	COMP
UPC	UpC	URIDYLYL (3'-5') CYTIDINE	COMP
UPN	RPN	UMGEKEHRTE POLNISCHE NOTATION	ELECT
UPS		ULTRAVIOLET PHOTOELECTRON SPECTROSCOPY	METH
UR	IR	ULTRAROT	METH

Abkürzung Akromym	Alternative	Bedeutung	Zuordnung
abbreviation akronym	alternative	meaning	related section
URF		UNASSIGNED READING FRAME(S)	POLYM
US		UNITED STATES	
USAEC		UNITED STATES ATOMIC ENERGY COMMISSION (US)	INSTI
USAN		UNITED STATES ADOPTED NAME	COMP
USDA		UNITED STATES DEPARTMENT OF AGRICULTURE (US)	INSTI
USEPA	EPA	UNITED STATES ENVIRONMENT PROTECTION AGENCY (US)	INSTI
USGEB		UNION DER SCHWEIZER.GESELLSCHAFT. FÜR EXPERIMENT.BIOLOGIE (CH)	ORGANIS
USM		ULTRA SOUND MICROSCOPE MICROSCOPY	INSTR
USNCI		UNITED STATES NATIONAL CANCER INSTITUTE	INSTI
USOS	U.S.O.S	UNITED STATES OCCUPATIONAL STANDARDS (US)	NORM
USP		UNITED STATES PHARMACOPOEIA (US)	NORM
USP		UNITED STATES PATENT	REG
USS		UNION SCHWEIZERISCHER SEIFENFABRIKANTEN (CH)	ORGANIS
UTI		URINARY TRACT INFECTION	CLINCHEM
UTP		URIDINE 5'-TRIPHOSPHATE	COMP
UVP		UMWELTVERTRÄGLICHKEITSPRÜFUNG (CH)	ENVIR REG
UVS		ULTRA VIOLET SPECTROSCOPY	METH
UVV		UNFALL VERHÜTUNGSVORSCHRIFTEN (D,BG)	REG
UVVIS	UV-VIS	ULTRA VIOLET VISIBLE SPECTROMETRY	METH
UWI		UMWELTWISSENSCHAFTLICHES INSTITUT (STUTTGART)	INSTI

Abkürzung Akromym	Alternative	Bedeutung	Zuordnung
abbreviation akronym	alternative	meaning	related section
VAA		VERBAND ANGESTELLTER AKADEMIKER (D)	ORGANIS
VAC		VOLTAGE ALTERNATING CURRENT	INSTR
VAL	Val	VALINE	COMP
VAS	VMA	VANILLIN MANDEL SÄURE	COMP
VAT	MWSt	VALUE ADDED TAX (US)	REG
VB		VALENCE BOND	THEOR
VBCHI	VBChI	VERBAND BASLER CHEMISCHER INDUSTRIELLER	ORGANIS
VBG		VERBAND DER DEUTSCHEN BERUFSGENOSSENSCHAFTEN (D)	ORGANIS
VBI		VERBAND BERATENDER INGENIEURE (D)	ORGANIS
VC		VINYL CHLORIDE	COMP
VCA		VIRAL CAPSID ANTIGENE	CLINCHEM
VCD		VIBRATIONAL CIRCULAR DICHROISM	METH
VCH		4-VINYL CYCLOHEXENE	COMP
VCH		VERBAND DES DEUTSCHEN CHEMIKALIEN GROß- UND AUßENHANDELS	ORGANIS
VCI		VERBAND DER CHEMISCHEN INDUSTRIE (D)	ORGANIS
VCI		VOLATILE CORROSION INHIBITOR	MATER
VCR		VIDEO CASSETTE RECORDER	INSTR
VDC		VINYLIDENE CHLORIDE	COMP
VDC	DC	VOLTAGE DIRECT CURRENT	ELECT
VDCh		VEREIN DEUTSCHER CHEMIKER (VORGAENGER DER GDCh,D)	ORGANIS
VDE		VERBAND DEUTSCHER ELEKTROTECHNIKER (D)	ORGANIS
VDEH	VDEh	VEREIN DEUTSCHER EISENHUETTEN- UND BERGLEUTE (D)	ORGANIS
VDG		VEREINIGUNG DEUTSCHER GEWÄSSERSCHUTZ (D)	ORGANIS
VDGH		VERBAND DER DIAGNOSTICA-& DIAGNOSTICAGERÄTE HERSTELLER	ORGANIS
VDI		VEREIN DEUTSCHER INGENIEURE (D)	ORGANIS
VDLUFA		VERB.DEUT.LANDWIRTSCHAFTL. UNTERSUCHUNGS-UND FORSCHUNGSANSTALTEN(D)	ORGANIS
VDM		VERBAND DER MINERALFARBENINDUSTRIE eV	ORGANIS
VDMA		VEREIN DEUTSCHER MASCHINEN-UND ANLAGENBAUER (D)	ORGANIS

Abkürzung Akromym abbreviation akronym	Alternative alternative	Bedeutung meaning	Zuordnung related section
VDR		VOLTAGE DEPENDENT RESISTOR	ELECT
VDRI		VEREIN DEUTSCHER REVISIONS INGENIEURE E.V. (D)	ORGANIS
VDSI		VEREIN DEUTSCHER SICHERHEITSINGENIEURE (D)	ORGANIS
VDTÜV		VEREINIGUNG DEUTSCHER TECHNISCHER ÜBERWACHUNGSVEREINE (D)	ORGANIS
VDU		VIDEO DISPLAY UNIT	INSTR
VEPA		VERY HIGH EFFICIENY PARTICULATE AIR FILTER	TECHN
VES		VENTRIKULAERES EXTRASYSTOLEN	CLINCHEM
VEW		VOLL ENTSALZTES WASSER	TECHN
VF		VULCANIZED FIBRE	MATER
VFDB		VEREIN ZUR FÖRDERUNG DES DEUTSCHEN BRANDSCHUTZES (D)	ORGANIS
VFET		VARIABLE FIELD EFFECT TRANSISTOR	ELECT
VFFA		VOLATILE FREE FATTY ACID	COMP
VFWL	ASHEA	VEREIN ZUR FÖDRERUNG DER WASSER UND LUFTHYGIENE (CH)	ORGANIS
VG		VACUUM GENERATORS (GB)	COMPANY
VGB		TECHNISCHE VEREINIGUNG DER GROßKRAFTWERKSBETREIBER (D)	ORGANIS
VGCT		VEREIN FÜR GERBEREICHEMIE UND TECHNIK (D)	ORGANIS
VGL		SCHWEIZERISCHE VEREINIGUNG GEWÄSSERSCHUTZ UND LUFTHYGIENE (CH)	ORGANIS
VHCP		VERBOND VAN HANDELAREN VAN CHEMISCHE PRODUKTION (NL)	ORGANIS
VHH		VOLATILE HALOGENATED HYDROCARBONS	COMP
VHI		VAPOUR HAZARD INDEX	REG
VICI		VERBAND DER INGENIEURE DER CHEMISCHEN INDUSTRIE	ORGANIS
VIP		VASOACTIVE INTESTINAL POLYPEPTID	CLINCHEM
VIP		VISUAL IMAGE PROCESSING	ELECT
VIPS		VEREINIGUNG D. IMPORTEURE PHARMAZEUTISCHER SPEZIALITÄTEN (CH)	ORGANIS
VIS		VISIBLE SPECTROSCOPY	METH
VKE		VERBAND KUNSTSTOFFERZEUGENDE INDUSTRIE (D)	ORGANIS
VKI		VERBAND DER KÖRPERPFLEGEMTTELINDUSTRIE	ORGANIS

Abkürzung Akromym abbreviation akronym	Alternative alternative	Bedeutung meaning	Zuordnung related section
VLDL		VERY LOW DENSITY LIPOPROTEIN = BLOOD PROTEIN	CLINCHEM
VLF		VERY LOW FREQUENCY	INSTR
VLGF		VERY LARGE GENE FRAGMENTS	CLINCHEM
VLM		VINYLCHLORIDE MONOMER	COMP POLYM
VLS		VALENCE LEVEL SPECTROSCOPY	METH
VLSI		VERY LARGE SCALE INTEGRATION	ELECT METH
VM		VALINOMYCIN	COMP
VMA		VANILLOMANDELIC ACID (=DL-4-HYDROXY-3-METHOXY MANDELIC ACID)	COMP
VMP		VARIABLE MAJOR PROTEIN	CLINCHEM
VNCI		VEREINIGUNG VAN DER NEDERLANDSE CHEMISCHE INDUSTRIE (NL)	ORGANIS
VNO		VERBAND DER NEDERLANDSE ONDERNEMIGEN (NL)	ORGANIS
VOC	VOC-	VINYL OXY CARBONYL-LIGAND	COMP
VOC		VOLATILE ORGANIC CARBON	COMP
VOCCL	VOCCl	VINYL CHLORO FORMATE	COMP
VOTC		VEREIN ÖSTERREICHISCHER TEXTIL CHEMIKER (A)	ORGANIS
VP1		VIRALES PROTEIN 1	CLINCHEM
VPE		VERNETZTES POLYETHYLEN	MATER POLYM
VR		VOLTAGE REGULATION	INSTR
VRT		VIBRATION-ROTATIONAL TRANSISTIONS	EVAL
VSA		VIRUS CAPSID ANTIGEN	CLINCHEM
VSA		VEREIN SCHWEIZERISCHER ABWASSERFACHLEUTE (CH)	ORGANIS
VSEPR		VALENCE SHELL ELECTRON PAIR REPULSION	THEOR
VSG		VARIABLE SURFACE GLYCOPROTEIN	CLINCHEM
VSLF		VERBAND SCHWEIZERISCHER LACK- UND FARBENFARBRIKANTEN (CH)	ORGANIS
VSS		VASOSPASTISCHES SYNDROM	CLINCHEM
VTC		VINYL TRICHLOROSILANE	COMP
VTCC		VEREIN DER TEXTILCHEMIKER UND COLORISTEN (CH)	ORGANIS
VTE		VARIABLE TIME EXPANSION	EVAL INSTR
VTEO		VINYL TRIETHOXYSILANE	COMP
VTF		VOLTAGE TUNABLE FILTER	INSTR
VTG		VERFAHRENSTECHNISCHE GESELLSCHAFT	ORGANIS
VTL		VARIABLE THRESHOLD LOGIC	ELECT

Abkürzung Akromym	Alternative	Bedeutung	Zuordnung
abbreviation akronym	alternative	meaning	related section
VTMO		VINYLTRIMETOXYSILANE	COMP
VTMOEO		VINYL-TRIS-(2-METHOXYETHOXY)-SILANE	COMP
VTR		VIDEO TAPE RECORDER	INSTR
VUV		VACUUM ULTRA VIOLET	INSTR
VUVS		VACUUM ULTRA VIOLET SPECTROSCOPY	METH
VVC		VOLTAGE VARIABLE CAPACITANCE	ELECT
VVO		VULCANIZED VEGETABLE OIL	MATER POLYM
VWD		VARIABLE WAVELENGTH DETECTOR	INSTR
VWF		VERORDNUNG WASSERGEFÄHRDENDER FLÜSSIGKEITEN (CH)	REG
VWF		VERBAND DER WISSENSCHAFTLER AN FORSCHUNGSANSTALTEN	ORGANIS
VZ		VERSEIFUNGSZAHL	METH
VÖCh	VöCh	VEREIN ÖSTERREICHISCHER CHEMIKER (A)	ORGANIS

Abkürzung Akromym abbreviation akronym	Alternative alternative	Bedeutung meaning	Zuordnung related section
WAA		WIEDERAUFARBEITUNGSANLAGE (D)	TECHN
WABOLU		INSTITUT FÜR WASSER-,BODEN-UND LUFTHYGIENE (BGA BERLIN,D)	INSTI
WACBDP		WIDE ANGLE CONVERGENT BEAM DIFFRACTION PATTERN	METH
WAS		WIRKSAMKEIT WASCHAKTIVER SUBSTANZEN	MATER
WAW		WIEDERAUFARBEITUNGSANLAGE WACKERSDORF (D)	TECHN
WAX		WEAK ANION EXCHANGER	MATER
WB		WEAK BEAM	INSTR
WBC		WHOLE BLOOD SAMPLE LEUCOCYTE COUNT	CLINCHEM
WBDF		WEAK BEAM DARK FILED METHOD	METH
WBGT		WET BULB GLOBE TEMPERATURE	CLINCHEM
WBT		WHITE BEAM TOPOGRAPHY	METH
WCOT		WALLCOATED OPEN TUBULAR	INSTR
WCS		WRITEABLE CONTROL STORE	ELECT
WCX		WEAK CATION EXHANGER	MATER
WD		WAVELENGTH DISPERSION	METH
WDK	WdK	WIRTSCHAFTSVERBAND DER DEUTSCHEN KAUTSCHUKINDUSTRIE eV	ORGANIS
WDS		WAVELENGTH DISPERSIVE SPECTROMETER	INSTR
WDX		WAVELENGTH DISPERSIV X-RAY ANALYSIS	METH
WE		WORKING ELECTRODE	INSTR
WERC		WORLD ENVIRONMENT AND RESOURCES COUNCIL (UN)	INSTI
WFA		WIEN FILTER ANALYSER	INSTR
WFC		WORLD FOOD COUNCIL (UN)	INSTI
WFCC		WORLD FEDERATION OF CULTURE COLLECTIONS (INT)	ORGANIS
WFP		WARENSCHUTZ FERTIG PRÄPARATE	MATER REG
WG		WEICHGLAS	MATER
WG	LWL	WAVE GUIDE TUBES	INSTR
WGA		WHEAT GERMAGGLUTIN	CLINCHEM
WGK		WASSERGEFÄHRDUNGSKLASSEN (CH)	REG
WGU		WISSENSCHAFTLICHE GESELLSCHAFT FÜR UMWELTSCHUTZ eV (AACHEN,D)	ORGANIS
WHG		WASSERHAUSHALTSGESETZ (D)	REG
WHO	OMS	WORLD HEALTH ORGANIZATION (UN , INT)	INSTI

Abkürzung Akromym	Alternative	Bedeutung	Zuordnung
abbreviation akronym	alternative	meaning	related section
WICEM		WORLD INDUSTRY CONFERENCE ON ENVIRONMENTAL MANGAGEMENT (UN)	CONF
WLD	TCD	WAERMELEITFAEHIGKEITSDETEKTOR	INSTR
WLN		WISWESSER LINE NOTATION (LIT)	LIT
WMO		WORLD METEROLOGICAL ORGANIZATION (UN)	INSTI
WORM		WRITE ONCE-READ MOSTLY	ELECT
WPAC		WORKING PARTY ON ANALYTICAL CHEMISTRY (FECS)	ORGANIS
WPCF		WATER POLLUTION CONTROL FEDERATION (US)	ORGANIS
WPEB		WORKING PARTY ON EDUCATION IN BIOTECHNOLOGY	ORGANIS
WPFC		WORKING PARTY ON FOOD CHEMISTRY (INT,FECS)	ORGANIS
WPO		WEAK PHASE OBJECT	MATER
WRC		WATER RETENTION CAPACITY	MATER
WSCOT		WALL COATED SUPER CAPACITY OPEN TUBULAR COLUMN	MATER
WSV		WASSER- UND SCHIFFAHRTSVERWALTUNG (D)	INSTI
WTR		WORKING GROUP TOXICOLOGY OF RUBBER CHEMICALS	ORGANIS
WVA		WIRTSCHAFTSVERBAND ASBEST (D)	ORGANIS
WVLI		WIRT. VEREINIGUNG DER LEBENSMITTELINDUSTRIE (BERLIN,D)	ORGANIS
WWF		WORLD WILDLIFE FUND (INT)	ORGANIS

Abkürzung Akromym / abbreviation akronym	Alternative / alternative	Bedeutung / meaning	Zuordnung / related section
XAES		X-RAY INDUCED AUGER ELECTRON SPECTROSCOPY	METH
XDC		X-RAY DOUBLE CRYSTAL DIFFRACTION	METH
XDP		XANTHOSINE-5'-DIPHOSPHATE	COMP
XEAPS		X-RAY INDUCED ELECTRON APPEARANCE SPECTROSCOPY	METH
XED		X-RAY ENERGY DISPERSIVE DIFFRACTOMETER	INSTR
XEM		EXO-ELECTRON MICROSCOPY	METH
XES		EXO ELECTRON SPECTROSCOPY	METH
XFA	XRFA,RFA	X-RAY FLUORESCENCE ANALYSIS	METH
XFLU	XRF	X-RAY FLUORESCENCE	INSTR
XIN		X-RAY INTERFEROMETER	INSTR
XLD		XYLOSE-LYSIN DESOXYCHOLATE	CLINCHEM
XLPE		VERY LARGE POLYMERIZED ETHYLENE	MATER POLYM
XMP		XANTHOSINE-5'-MONOPHOSPHATE	COMP
XMPA		X-RAY ELECTRON MICROPROBE ANALYSIS	METH
XNBR		ACRYLNITRILE-BUTADIENE RUBBER	MATER POLYM
XPS	ESCA	X-RAY PHOTO ELECTRON SPECTROSCOPY	METH
XRD		X-RAY DIFFRACTION	METH
XRF		X-RAY FLUORESCENCE	METH
XRFA	XFA	X-RAY FLUORESCENCE ANALYSIS	METH
XRITC		HETEROCYCLIC SUBSTUTUTED TETRAMETHYLRHODANINE ISOTHIOCYANATE	COMP
XSAD		X-RAY SMALL ANGLE DIFFRACTION	METH
XTP		XANTHOSINE-5'-TRIPHOSPHATE	COMP
XUV		WAVELENTH AT 0.5nm SYNCHROTRON RADIATION	INSTR

Abkürzung Akromym	Alternative	Bedeutung	Zuordnung
abbreviation akronym	alternative	meaning	related section
YAG		YTTRIUM ALUMINIUM GARNET (GRANAT)	MATER
YAP		YTTRIUM ALUMINIUM RPEROWSKIT	MATER
YIG		YTTRIUM IRON GARNET (GRANAT)	MATER
Z	Z- CBz-	CARBOBENZOXY LIGAND	COMP
ZAA		ZEEMANN ATOMIC ABSORPTION	METH
ZAAS		ZEEMANN ATOMIC ABSORPTION SPECTROSCOPY	METH
ZAB		ZERO ALPHA-BETA ABERRATION	EVAL
ZAED		ZENTRALSTELLE FÜR ATOMKERNENERGIE DOKUMENTATION (D)	LIT
ZAF		CORRELATION FOR ATOMIC NUMBER Z, ABSORPTION AND FLUORESCENCE	METH
ZAP		ZONE AXIS PATTERN	METH
ZAV		ZENTRALSTELLE FÜR ARBEITSVERMITTLUNG (D)	INSTI
ZBDC		ZINC DIBENZYLDITHIOCARBAMATE	COMP
ZBKM		ZENTRALBÜRO FÜR KERNMESSUNGEN (EG)	INSTI
ZBPD		ZINC DIBUTYL DIPHOSPHATE	COMP
ZBS		ZERO BIAS SCHOTTKY DIODE	INSTR ELECT
ZBX		ZINC DIBUTYL-XANTHATE	COMP
ZDBC		ZINC DIBUTYL DITHIOCARBAMATE	COMP
ZDEC		ZINC DIETHYLDITHIOCARBAMATE	COMP
ZDMC		ZINC DIMETHYLDITHIOCARBAMATE	COMP
ZDO		ZERO DIFFERENTIAL OVERLAP	THEOR
ZDV		ZERO DEAD VOLUME	EVAL
ZEBS		ZENTRALE ERFASSUNGS- UND BEWERTUNGSSTELLE FÜR UMWELTCHEMIKALIEN (BGA)	ENVIR LIT

Abkürzung Akromym	Alternative	Bedeutung	Zuordnung
abbreviation akronym	alternative	meaning	related section
ZEH		ZINC ETHYL HEXANOATE	COMP
ZEPC		ZINC ETHYL-PHENYL DITHIOCARBONATE	COMP
ZFA	ZfA	ZENTRALSTELLE FÜR ABFALLBESEITIGUNG (BGA)	INSTI
ZFP	ZfP	ZERSTÖRUNGSFREIE PRÜFUNG	MATER METH
ZGS		ZIRCONIUM GRAIN STABILIZED	MATER
ZIA		ZONE IMMONO ASSAY	METH
ZIAS		ZENTRALINSTITUT FÜR ARBEITSSCHUTZ (DRESDEN,DDR)	INSTI
ZID		ZENTRALSTELLE FÜR INFORMATION UND DOKUMENTATION (EG)	INSTI
ZIGE		ZENTRALE INFORMATIONSTELLE ZUR GENEHMIGUNGSVERGABE & EMISSIONSKATASTER (NRW,D)	INSTI
ZINEB	Zineb	ZINC ETHYLENE-BIS(DITHIOCARBAMATE)	COMP
ZIRAM	Ziram	BIS (DIMETHYLCARBAMADITHIATO-S,S') ZINC	COMP
ZIX		ZINC ISOPROPYL XANTHATE	COMP
ZKBS		ZENTRALE KOMMISSION FÜR BIOLOGISCHE SICHERHEIT	INSTI
ZKR		ZENTRALKOMMISSION FÜR DIE RHEINSCHIFFART (STRAßBURG)	INSTI
ZLM		ZENTRALLABOR FÜR MUTAGENITÄTSPRÜFUNG (FREIBURG ,DFG)	INSTI
ZMBH		ZENTRUM FÜR MOLEKULARE BIOLOGIE HEIDELBERG (D)	INSTI
ZMBI		ZINC-2-MERCAPTO BENZIMIDAZOLE	COMP
ZMBT		ZINC-2-MERCAPTO-BENZTHIAZOLE	COMP
ZNS	CNS	ZENTRALES NERVEN SYSTEM	CLINCHEM
ZOLZ		ZERO ORDER LAUE ZONE	EVAL
ZPCK		N-CBZ-L-PHENYLALANINE CHLOROMETHYL KETONE	COMP
ZPE		ZERO POINT ENERGY	THEOR
ZPT		BIS-(N-OXO PYRIDINE-2-THIONATO) ZINC	COMP
ZTDI		ZENTRALSTELLE FÜR TEXTILDOKUMENTATION UND INFORMATION (D)	INSTI
ZVEI		ZENTRALVERBAND DER ELEKTROTECHNISCHEN INDUSTRIE (FRANKFURT,D)	ORGANIS

LÖSUNGSMITTEL

SUPPLEMENT PILZE
1 × jährlich
Abonnement DM 15,-
pro Ausgabe

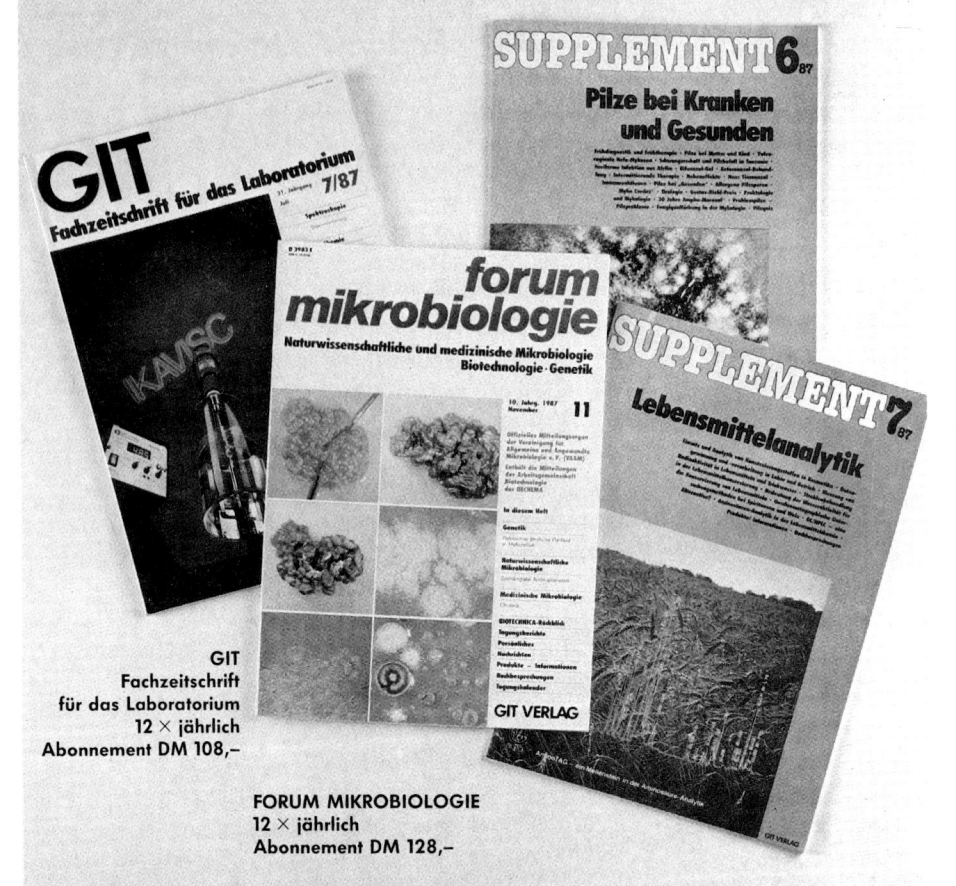

GIT
Fachzeitschrift
für das Laboratorium
12 × jährlich
Abonnement DM 108,-

FORUM MIKROBIOLOGIE
12 × jährlich
Abonnement DM 128,-

SUPPLEMENT LEBENSMITTEL
1 × jährlich
Abonnement DM 15,-
pro Ausgabe

**Für Praktiker – anwendungsbezogene Fachzeitschriften aus dem
GIT VERLAG**